Fabrication of Graphene from Camphor

Fabrication of Graphene from Camphor

Emerging Energy Applications

Harsh A. Chaliyawala, Kashinath Lellala,
Govind Gupta and Indrajit Mukhopadhyay

CRC Press is an imprint of the
Taylor & Francis Group, an **informa** business

First edition published 2021 by
CRC Press
6000 Broken Sound Parkway NW, Suite 300, Boca Raton, FL 33487-2742

and by CRC Press
2 Park Square, Milton Park, Abingdon, Oxon, OX14 4RN

CRC Press is an imprint of Taylor & Francis Group, LLC

Library of Congress Cataloging-in-Publication Data

Library of Congress Control Number: 2021933609

ISBN: 978-0-367-67723-7 (hbk)
ISBN: 978-0-367-68638-3 (pbk)
ISBN: 978-1-003-13837-2 (ebk)

Typeset in Times
by Deanta Global Publishing Services, Chennai, India

Contents

Preface

We have been investigating the optoelectronic applications of graphene since 2016 because we saw that it has the potential to significantly improve human lives in several ways such as energy harvesting, energy storage, optical communication and sensors. There is a sizable literature on graphene synthesis and properties and some on devices and applications. However, only some provide in-depth discussions about the many possible ways of graphene synthesis using a solid source of carbon in combination with nanostructures for optical devices. This less explored material thus gave us the idea to bring out *Fabrication of Graphene from Camphor: Emerging Energy Applications.*

A large group of emerging materials and devices is being extensively studied to replace silicon due to its scaling limit. Germanium has been substituted by silicon roughly half a century ago by moving up one block on group IV of the periodic table. Interestingly, moving up one more block, we reach carbon, which is widely used as a substitute for next-generation electronics due to its impressive crystal structures, or allotropes. As we near the end of Moore's law with silicon-based technology, graphene, among carbon allotropes, has the potential to become the next candidate material for future technology and could help move electronics beyond Moore's law. The discovery of graphene, a single atomic layer of carbon sheet, in 2004 has prompted research into its potential in developing future electronic and photonic devices owing to its exceptional electronic, photonic, and mechanical properties. The various methods involved in the fabrication of graphene are still in their early stages, and engineers need to devise methods for mass production of large, uniform sheets of pure, single-planed graphene sheets. The development of ultra-fast-growing technologies mostly relies on the fundamental understanding of novel 2D materials with unique properties as well as new designs of device architectures with more diverse and better functionalities.

Over the past few decades, there have been tremendous innovations in the development of graphene sheets for optoelectronics and photonics. The term "photonics" is concerned with light (photons or particles of light), and it deals with its properties, generation, manipulation and detection. "Optoelectronics" refers to the electronic effects induced by light or devices relying on such working principles. It is estimated that more than 1 trillion semiconductor devices were shipped for the first time in 2016 due to a rising demand from new application areas such as the Internet of Things, wearable electronics, smart sensors and robotics. Graphene is a zero-bandgap semimetal that has attracted significant attention due to its strong interaction with photons of a wide energy range, from microwave to ultraviolet, as well as its high carrier mobility for high-speed applications in a broad wavelength range. Graphene has been extensively investigated as a promising material for various types of high-performance sensors due to its large surface-to-volume ratio, remarkably high carrier mobility, high carrier density, high thermal conductivity, extremely high mechanical strength and high signal-to-noise ratio. Graphene is also a promising material for energy storage systems such as high energy density batteries due to its maximum surface-to-bulk ratio along with its exceptional electrical and thermal transport properties.

This book is organized on the basis of the different applications presented. This will allow readers, especially seasoned investigators who have a practical problem in mind, to jump to the relevant chapter quickly. While we have tried to include grapheme-based energy conversion and storage applications, this book is not meant to be exhaustive and we encourage interested readers to peruse other literature. The book can be used by students of electrical and electronics engineering, applied physics and materials science with a focus on next-generation low dimension materials such as graphene. The content of this book forms the basis to understand the unique nanostructures in combination with camphor-based graphene sheets, which will be very useful for practicing engineers in the field of electronics device design and technology, solid-state physics and semiconductors, nanoscience and nanotechnology, and nanophotonic and nanoelectronic devices. The first chapter is an introductory chapter which discusses the discovery, processing methods, optical properties of graphene and nanowire Schottky junction for energy applications. The subsequent chapters will discuss about the fabrication of graphene by camphor and its application in photodetectors and as an anodic material for energy storage applications. The final chapter highlights the latest developments' status, perspectives and our views on advancement of graphene. We thank the publisher for taking us on this journey. We are grateful to our colleagues for their wonderful input, both directly as contributors in this book and indirectly through discussions and suggestions. It has been a great pleasure working with them, and we trust they feel the same, too. Finally, we thank our family for their unwavering support.

Harsh Chaliyawala
Kashinath Lellala
Govind Gupta
Indrajit Mukhopadhyay
Winter 2021

Author Bios

Dr. Harsh A. Chaliyawala holds a PhD in Physics from Solar Research and Development Centre, Pandit Deendayal Petroleum University, Gandhinagar, Gujarat, India. He is currently Teaching Assistant of the Department of Physics at Sardar Vallbhbhai National Institute of Technology, Surat, India. His research interests include low dimensional systems, nanostructures, solid-state physics and semiconductor physics. He has published more than 15 peer-reviewed papers with a total citing of 140 and an h index of 6 and an i10 index of 5. He has been awarded a Senior Research Fellowship from the Council of Scientific & Industrial Research (CSIR), India. He is also a reviewer of various international journals of Institute of Physics (IOP), American Institute of Physics (AIP) and Elsevier publishers. He has also received the IOP outstanding reviewer award in 2019.

Dr. Kashinath Lellala obtained his PhD in Materials Science in 2018 from the University of Mysore, India. His current research focuses mainly on the fabrication of hybrid heterostructure nanocomposites (metal oxide/metal sulfides) and graphene-based composites for energy and environmental applications including lithium-ion batteries, fuel cells, water remediation, and water splitting. He has served as Lecturer in Physics at Iringa University, Tanzania (2009) and Royal University of Bhutan, India. He has also visited the University of South Australia as a visiting researcher during his PhD. He has also worked as a research associate at the Department of Solar Energy, Pandit Deendayal Petroleum University, and is presently working as a post-doctoral fellow at Division of Materials Science, Department of Engineering Science and Mathematics, Lulea University of Technology, Lulea, Sweden. He has published more than 25 papers with a total citation of 314 and an h index of 11 and an i10 index of 11.

Dr. Govind Gupta has been Senior Principal Scientist and Head, Sensor Devices & Metrology group of CSIR-National Physical Laboratory, Delhi, India. He is also Professor at Academy of Scientific & Innovative Research (AcSIR), Delhi. His area of expertise is the thin film growth and fabrication of state-of-the-art optoelectronics and gas sensing devices. Dr. Gupta has made significant contributions in the field of III-V metal oxide semiconductor heterojunctions, which are essentially beneficial for the fabrication of next-generation optoelectronic devices. He is Associate Academician of Asia Pacific Academy of Materials (APAM) and a recipient of the prestigious Materials Research Society of India (MRSI) medal for his significant contribution to Materials Science & Technology. He is also a recipient of Young Scientist Medal (Platinum Jubilee Award) from National Academy of Sciences (NASI), India (2010), Outstanding Young Scientist of the Year Award (OYSA) from CSIR-National Physical Laboratory, India (2009), Young Scientist Award by IFW-ICNN, India (2009), BOYSCAST fellowship by Department of Science & Technology, India (2007) and Young Scientist Award (YSA) from Ministry of Science & Technology, Government of India. Dr. Gupta is a member of many Scientific Societies including MRS (USA), ACS (USA), MRSI, SSI, EMSI, MSI, Vijyana Bharti, IAAM, etc., and a recognized panel reviewer for many scientific journals. He has published more than 240 research publications with ~4000 citations.

Dr. Indrajit Mukhopadhyay has been Professor in the Department of Solar Energy, Pandit Deendayal Petroleum University, Gandhinagar, Gujarat, India. He works in the field of photovoltaic and photoelectrochemical solar cells. He established the Solar Research and Development Center (SRDC) at PDPU that houses the state-of-the-art facilities to carry out cutting-edge experimental research on solar energy conversion and galvanic energy storage application. Dr. Mukhopadhyay received his PhD. from IIT, Mumbai. He has published more than 110 research publications in journals of international repute with a total citation of 2187 with an h index of 25 and an i10 index of 53. He is engaged in active collaboration with prestigious institutes/universities in Germany, Japan, Israel, Republic of Korea, France, Czech Republic and the USA. His recent interest is in developing large-area graphene for optoelectronic applications.

1 Introduction

Harsh A. Chaliyawala, Kashinath Lellala, Govind Gupta and Indrajit Mukhopadhyay

Contents

Since its invention in 2004, graphene, a novel 2-dimensional (2D) material, has shown its uniqueness in exhibiting extraordinary properties. Graphene is a single layer of carbon atoms (C-C distance of 0.142 nm) with a hexagonal closed pack structure. It is an ultrathin, mechanically strong, transparent and flexible conducting material. The electrical conductivity of graphene is 1.4 times higher than that of Cu or Si (conductivity of graphene is ~80×10^6 Sm^{-1}) and also has high thermal conductivity (Graphene: 3–5 $KWm^{-1}K^{-1}$, Cu: $400Wm^{-1}K^{-1}$), making it the best thermal conductor [1]. Its conductivity can be increased over a large range either by changing the number of layers of graphene, also known as chemical doping, or by applying electric fields. Moreover, it also has high electron mobility (15,000 $cm^2/V.s$) and a very large specific surface area (SSA ~ 2,630 m^2/g) that render the material several interesting properties for various optoelectronic applications [2]. Further, graphene sheets are flexible as well as chemically inert, giving it a dual role: as an electrode and as a protective layer. However, some concerns associated with its high transparency (absorbs 2.3%), which is not favorable for solar cell applications, need to be resolved. This problem can be ideally solved by doping graphene to make *p*- to *n*-type. Together, these extraordinary properties of graphene make it an excellent candidate for energy harvesting devices such as solar cells as well as for sensors, photodetectors, etc.

Motivated by all the superior properties of graphene, different processes are currently underway to achieve large-area graphene sheets with low sheet resistance and high transmittance values [3], [4]. In order to obtain the above-mentioned properties, researchers are employing various synthesis methods such as mechanical cleaving (exfoliation) [5], chemical exfoliation [6], chemical synthesis [7], and chemical vapor deposition (CVD) [8]. A few other techniques such as unzipping nanotube [9], [10] and microwave irradiation assisted synthesis [11] are also reported elsewhere. Among all the synthesis processes, chemical vapor deposition (CVD) is one of the most versatile techniques to synthesize large-area graphene with a controllable number of layers and single-crystal domains. In this technique, the carbon precursors are decomposed on Cu surfaces due to their low carbon solubility and high self-limiting factor [12]–[14]. Recently, gaseous hydrocarbons and liquid carbon sources have been replaced with solid carbon sources such as camphor ($C_{10}H_{16}O$) and polystyrene using a very facile and low-cost approach to avoid the toxicity of gaseous sources and to overcome the challenge of developing large-area graphene on Cu foil with optimum optoelectronic properties [15]–[17]. Among a few researchers working on this area, Kalita et al. have initiated the growth of mono/bi-layer graphene by employing camphor as a carbon source on different metal substrates [17], [18].

Combining graphene with metals, semiconductors, ceramics, polymers and biomaterials may have a great influence on the properties of the material. A metal/semiconductor Schottky junction can be formed when graphene is placed close to semiconductors such as silicon (Si). As compared to its bulk and thin-film structures, vertical Si nanowire arrays (SiNWAs), with its large surface to volume ratio, strong light trapping ability and superior electron transport properties, have shown a tremendous potential as a building block for various optoelectronic devices. Graphene-based semiconductor heterojunctions have been widely used in electronic and optoelectronic devices, such as field-effect transistors (FETs), photodetectors (PDs), photovoltaics, lasers and light-emitting diodes [19], [20], [21]. Among various operational nanoscale devices, SiNWAs based photodetectors (PDs) have received special research attention due to their ability to probe incoming infrared light with high sensitivity and their excellent photo response, which can be a future component for large scale applications in military surveillance, target detection and tracking [19]. Graphene/Si Schottky junction–based photodetectors inherit a lot of merits as they permit a simple device architecture, self-driven operation and broadband photodetection. In order to exhibit the extraordinary properties of both the materials, recent research interests are to combine vertical SiNWAs with a low-cost synthesized graphene sheet, to construct high-performance energy selective metal/semiconductor Schottky junction photodetectors.

Another area of potential application for an energy-efficient device is rechargeable batteries. Recently, rechargeable lithium-ion batteries (LIBs) have attracted huge attention as a great power source for portable electronic devices and electric vehicles due to their high power density, higher energy, long life cycle, non-toxicity and flexible, easy fabrication for energy devices, and therefore can replace major gasoline-powered transportation and can in turn reduce the greenhouse gas emission [22]. Interestingly, graphitic-based materials have been commercially used as anode in LIBs due to its interesting properties which supports moderate energy density and stable cycle performance [23]. Graphene has shown remarkable and excellent cyclic performance, high energy and power density in LIBs due to its fascinating physiochemical properties. Graphene's intrinsic and extrinsic properties can be altered by constructing different types of layered sheets and defect-free carbon structuring for high energy density and better cycle performance of Li ion battery. Developing a new class of graphene or graphene-based materials for energy applications has attracted the attention of researchers. Thus, fabrication of high-performance and higher-efficiency anodic materials is needed for the advancement in lithium-ion batteries with high energy density for all electronic and electrical devices [24–25].

In this chapter, we have reviewed the journey of camphor-based graphene synthesis and its highly efficient energy applications. In Chapter 2, we will briefly describe the synthesis of graphene sheets from various carbon sources. Gas, liquid and solid carbon precursors are discussed to understand the mechanism for the synthesis of graphene sheets. In Chapter 3, we will elaborate the nucleation and growth mechanisms followed by the analysis of various structural, morphological, and optical properties of camphor-based graphene. Chapter 4 will focus on the near infrared (NIR) photodetectors based on Gr and SiNWAs Schottky junction, which elucidates a brief understanding about the electrical performance parameters, effect of nanowire length on the device performance and self-driven operation of fabricated NIR photodetectors (PDs). Later, the potential applications of Gr-based high energy density batteries are presented in Chapter 5, followed by the Current and Future Perspective in Chapter 6.

REFERENCES

1. T. H. E. Royal, S. Academy, and O. F. Sciences, "Compiled by the class for physics of the Royal Swedish Academy of Sciences Graphene," vol. 50005, no. October, pp. 0–10, 2010.
2. A. A. Balandin, S. Ghosh, W. Bao, I. Calizo, D. Teweldebrhan, F. Miao, and C. N. Lau, "Superior thermal conductivity of single-layer graphene," *Nano Lett.*, vol. 8, no. 3, pp. 902–907, 2008.
3. N. O. Weiss, H. Zhou, L. Liao, Y. Liu, S. Jiang, and Y. Huang, "Graphene: an emerging electronic material," *Wiley Adv. Mater.*, vol. 24, no. 43, pp. 5782–5825, 2012. https://doi.org/10.1002/adma.201201482
4. J. Jiang, J. Wang, and B. Li, "Young's modulus of graphene: a molecular dynamics study," *Phys. Rev. B*, vol. 125, p. 175110, 2019. https://doi.org/10.1063/1.5091753

5. K. S. Novoselov, et al., "Electric field effect in atomically thin carbon films," *Science*, vol. 306, no. 5696, pp. 666–669, Oct. 2004.
6. M. J. Allen, V. C. Tung, and R. B. Kaner, "Honeycomb carbon: a review of graphene what is graphene ?," *Chem. Rev.*, vol. 110, pp. 132–145, 2010.
7. S. Park, and R. S. Ruoff, "Chemical methods for the production of graphenes," *Nat. Nanotechnol.*, vol. 4, no. 4, pp. 217–224, Apr. 2009.
8. A. Reina, et al., "Large area, few-layer graphene films on arbitrary substrates by chemical vapor deposition," *Nano Lett.*, vol. 9, no. 1, pp. 30–35, Jan. 2009.
9. L. Jiao, L. Zhang, X. Wang, G. Diankov, and H. Dai, "Narrow graphene nanoribbons from carbon nanotubes," *Nature*, vol. 458, no. 7240, pp. 877–880, Apr. 2009.
10. D. V. Kosynkin, et al., "Longitudinal unzipping of carbon nanotubes to form graphene nanoribbons," *Nature*, vol. 458, no. 7240, pp. 872–876, Apr. 2009.
11. G. Xin, W. Hwang, N. Kim, S. M. Cho, and H. Chae, "A graphene sheet exfoliated with microwave irradiation and interlinked by carbon nanotubes for high-performance transparent flexible electrodes," *Nanotechnology*, vol. 21, no. 40, 2010.
12. F. H. L. Koppens, T. Mueller, P. Avouris, A. C. Ferrari, M. S. Vitiello, and M. Polini, "Photodetectors based on graphene, other two-dimensional materials and hybrid systems," *Nat. Nanotechnol.*, vol. 9, no. 10, pp. 780–793, 2014.
13. M. Amirmazlaghani, F. Raissi, O. Habibpour, J. Vukusic, and J. Stake, "Graphene-Si schottky IR detector," *IEEE J. Quantum Electron.*, vol. 49, no. 7, pp. 589–594, 2013.
14. H. K. Raut, V. A. Ganesh, A. S. Nair, and S. Ramakrishna, "Anti-reflective coatings: a critical, in-depth review," *Energy Environ. Sci.*, vol. 4, no. 10, pp. 3779–3804, 2011.
15. T. Liang, Y. Kong, H. Chen, and M. Xu, "From solid carbon sources to graphene," *Chinese J. Chem.*, vol. 34, no. 1, pp. 32–40, 2016.
16. R. Papon, G. Kalita, S. Sharma, S. M. Shinde, R. Vishwakarma, and M. Tanemura, "Controlling single and few-layer graphene crystals growth in a solid carbon source based chemical vapor deposition," *Appl. Phys. Lett.*, vol. 105, no. 13, pp. 133103–133108, 2014.
17. G. Kalita, K. Wakita, and M. Umeno, "Monolayer graphene from a green solid precursor," *Phys. E Low-Dimensional Syst. Nanostructures*, vol. 43, no. 8, pp. 1490–1493, 2011.
18. S. Sharma, G. Kalita, R. Hirano, Y. Hayashi, and M. Tanemura, "Influence of gas composition on the formation of graphene domain synthesized from camphor," *Mater. Lett.*, vol. 93, no. February 2013, pp. 258–262, 2013.
19. G. Fan, et al., "Graphene/silicon nanowire schottky junction for enhanced light harvesting," *ACS Appl. Mater. Interfaces*, vol. 3, no. 3, pp. 721–725, 2011.
20. L. B. Luo, et al., "Light trapping and surface plasmon enhanced high-performance NIR photodetector," *Sci. Rep.*, vol. 4, pp. 3914–3922, 2014.
21. X. Wang, Z. Cheng, K. Xu, H. K. Tsang, and J. Bin Xu, "High-responsivity graphene/silicon-heterostructure waveguide photodetectors," *Nat. Photonics*, vol. 7, no. 11, pp. 888–891, 2013.
22. D. Liu, and G. Cao, "Engineering nanostructured electrodes and fabrication of film electrodes for efficient lithium ion intercalation," *Energy Environ. Sci.*, vol. 3, no. 9, pp. 1218e1237, 2010.

23. S. Wu, R. Xu, M. Lu, R. Ge, J. Iocozzia, C. Han, B. Jiang, and Z. Lin, "Graphene-containing nanomaterials for lithium-ion batteries," *Adv. Energy Mater.*, vol. 5, no. 21, pp. 1500400, 2015.
24. D. R. Dreyer, R. S. Ruoff, and C. W. Bielawski, "From conception to realization: an historical account of graphene and some perspectives for its future," *Angew Chem. Int. Ed.*, vol. 51, pp. 7640–7654, 2012.
25. K. S. Novoselov, V. I. Falko, L. Colombo, P. R. Gellert, M. G. Schwab, and K. Kim, "A roadmap for graphene," *Nature*, vol. 490, pp. 192–200, 2012.

2 Graphene sheets from various carbon precursors

Kashinath Lellala, Harsh A. Chaliyawala and Indrajit Mukhopadhyay

Contents

2.1 INTRODUCTION

Carbon is the most common element of life on Earth and it exists in many different allotropic forms, exhibiting various physicochemical properties. The best-known natural allotropes of carbon are graphite and diamond. After the discovery of graphene by Novoselov and Geim [1], there is a huge demand in the the field of condensed matter physics and material science. Graphene has obtained a new concept and invention/idea of physics and their potential applications [2]. Graphene is a mono atomic hexagonal layer of graphite with sp^2 hybridized carbon atoms forming a honeycomb-like structure in a two-dimensional (2D) crystal with unique features. With such remarkable physicochemical properties it surpasses other materials and its physical behavior imparts astounding high carrier mobility and higher charge carrier concentrations at lower temperature, special electronic structure and inconsistent quantum Hall effect [3–8]. As a novel class of material, graphene has excellent intrinsic and extrinsic mechanical [9], electronic, thermal [10], magnetic [11], and electrical [12] properties[13]. The current methods for generating graphene sheets are chemical vapor deposition (CVD) [14], solvothermal/hydrothermal synthesis [15], micro-mechanical exfoliation of graphite [16], epitaxial growth [17] and reduction of graphene oxide [18] which is briefly discussed in the later sections. Graphene and few-layer graphene sheets are grown by the CVD method using carbon-containing gases on a highly active catalytic metal surface or by surface segregation of carbon dissolved in metal surfaces. Depending upon the solubility of the carbon the growth dominant can be evaluated. For instance, graphite oxide was independently synthesized in the late 1950s by Hummers [19], in 1898 by Staudenmaier [20], and in 1859 by Brodie [21], and chemical reduction of graphene oxides was previously reported in 1962 [22]. The synthesis of monolayer graphene using silicon carbide as substrate was reported in 1975. [23]. The above synthesis methods did not yield graphene or graphitic materials with unique physicochemical properties but laid the foundation for developing such materials in later days [24]. Developing a large-area and scalable graphene synthesis is an alternative solution or approach to resolve needed solution. The current technique for synthesizing high quality and large area graphene employs the Hummers' method developed in the 1950s by William Hummers. [19]. In the year 2013, graphene grown via the CVD method was shown to exhibit 90% of the theoretical strength of pristine graphene, as demostrated by Lee et al. [25].

In this study, large-area graphene nanosheets were grown on Cu foil by using single-step CVD at high-temperature conditions and camphor as the solid carbon precursor. Single and multi-layer graphene nanosheets were developed by the abovementioned method, and they can be used for potential applications such as photodetector and lithium-ion batteries. Besides active research on graphene applications in recent years fundamental research on graphene has marked it as a promising candidate. The purpose of the present work is to review recent developments in the synthesis of graphene from different carbon sources. In general, methane is used as the best carbon precursor and Cu as the metal catalyst for nucleation at high temperatures of above 1000°C. Recently, Jang et al. synthesized monolayer graphene at 300°C using benzene as the precursor and Cu as the metal catalyst [26, 27]. This chapter mainly focuses on the synthesis of graphene using the CVD method and from carbon precursors in their various forms such as solids, liquids and gaseous materials.

2.2 VARIOUS METHODS OF GRAPHENE SYNTHESIS

2.2.1 Mechanical Exfoliation

Mechanical exfoliation of highly oriented pyrolytic graphite was carried out by Novoselov in early 2004 to obtain few-layer graphene. After fine-tuning, high-quality single-crystalline graphene sheets were obtained [4]. The most common exfoliation method is the scotch tape method, which utilizes an adhesive tape to pull graphene films off a graphite crystal, which is subsequently thinned down by rubbing further strips of the tape against the target substrate. This rather crude method creates a random array of single and double layers of graphene flakes on the target substrate and has been a key driver for investigating many properties of graphene. Since this method creates foldings in graphene, it cannot be reproduced with high accuracy, and therefore other mechanical and chemical exfoliation processes have been investigated. To address the difficulties of the scotch tape method, one group tried to exfoliate graphene from highly ordered pyrolytic graphite (HOPG) by using a sharp single-crystal diamond wedge to penetrate the graphite source. However, this method introduce a defect through shear stress after exfoliation.

Straightforward mechanical exfoliation methods are able to produce high-quality graphene flakes and are very beneficial for investigating the amazing characteristics of graphene, while liquid exfoliation (and reduction) methods are utilized for the production of transparent conducting oxides, conductive inks, and electrodes for Li-ion batteries and supercapacitors. Mechanical exfoliation, however, cannot be reliably scaled up to provide the reliable for optoelectronic applications placement and large-area high-quality graphene sheets desired for transistor and optoelectronic applications.

2.2.2 Liquid Phase Exfoliation of Graphite

The liquid phase exfoliation of graphite approach for the synthesis of graphene bypasses the oxidation/reduction process. The quality and quantity of graphene sheets synthesized depend on the surface energy of the solvent and the reductant. Making monolayer graphene is very difficult. Therefore, a specific solvent with high surface energy and water with an exfloiation agent such as N-methyl pyrrolidone is required. But this agent is very expensive and require special attention in handling. Next, sodium dodecylbenzene sulfonate can be used as surfactant in water for effective dispersion of graphite flakes and easy exfoliation can be achieved using the ultrasonication method or any other physical processes. This approach requires the use of polymers (synthetic/natural) for obtaining the graphitic carbon structure, and involves the following stages: (a) pyrolysis of polymers in the absence of oxygen at high temperatures and (b) liquid-phase exfoliation of the pyrolyzed material. Several liquid-phase exfoliation cycles are required to promote exfoliation performance. Exfoliation performance is the percentage by weight of the pyrolyzed material suspended in the solvent. For solid graphene sample preparation, the solvent is removed and the resulting residue is dried. Natural polymers used are biopolymers and their derivatives, for example polysaccharides (agricultural waste) such as chitosan and alginate. Synthetic polymers such as polystyrene polythiophene, poly-fur-based alcohol and polyacrylonitrile are widely used for the synthesis of graphene. In addition to synthetic polymers, solvents such as ionic liquids, water, acetone, methanol, NMP and DMF are used during exfoliation for the synthesis of graphene. This procedure can yield graphene sheets of dimensions between 1 micron and 100 microns from biopolymers. The graphene concentrations obtained in water suspensions may vary from 0.01 mg/ml to 0.2 mg/ml [27].

2.2.3 Epitaxial growth of graphene

A highly efficient technique for the production of high-quality and large-scale graphene films is the epitaxial growth of graphene. The search for a suitable substrate has so far been a major issue in the epitaxial growth of graphene since it imposes serious constraints on device architecture and functionality. Hence, there is an increasing interest in finding specific dielectrics that allow substrate-supported geometry. Ultrathin epitaxial graphene (EG) films grown on several substrates are promising candidates in the field of nanoelectronics. Growing large areas of EG on silicon carbide (SiC) is technologically feasible; moreover, SiC, being a semiconductor, is a convenient substrate for subsequent stages of device fabrication. Due to its atomically flat surface, almost free of dangling bonds and charge traps and its capability to adopt the electronic structure of graphene, hexagonal boron nitride (h-BN) is also an attractive substrate for nanodevices based on graphene [15–17]. The lattice constant of h-BN is closely equivalent to that of graphite having wide electrical bandgap and large optical phonon modes.

The epitaxial growth mechanisms of graphene at the atomic level and the effects of substrates by examining the transient stages at the growing edges of graphene have been well explained by Ozcelik et al. [11]. The formation of epitaxial graphene can be classified into two steps: the first step is nucleation and the next is graphene growth from the nucleated seed. The defects that form the step edges in atomic scale play a leading role in the process of nucleation at the substrate. The epitaxial growth of graphene on different substrates, such as Ru(0001), Pt(111), Ir(111), Cu(111), and Pd(111), was previously reported and well understood.

2.2.4 Chemical vapor deposition (CVD)

Besides the mechanical exfoliation and chemical reduction for the synthesis of graphene, chemical vapor deposition (CVD) on metal substrates is one of the best methods. CVD is widely used to fabricate bulk materials, composites and solid thin films of high purity. In general, the CVD process involves a precursor gas flow on a coated heated surface in a chamber and the chemical reactions near or on the hot surface leads to the deposition of a thin film or powder. The advantages of the CVD method are (i) bulk production of pure materials, (ii) good reproducibility and uniform film formation, (iii) controlled growth and nucleation and (iv) controlled surface morphology and growth orientations. Besides the advantages a few drawbacks of the CVD technique are (i) use of high temperature above 600°C, (ii) use of toxic and flammable gas, (iii) restrictions in developing multicomponent materials and growth and (iv) high cost [28]. Graphene could be prepared by the CVD method from decomposition of methane/acetylene/ethylene on metal surface. Addition of carbon atoms from methane gas is widely used for the synthesis of graphene via epitaxial growth in which carbon sources are attached to the surface of a metallic substrate. Another method is carbonization of waste materials or biomass. These methods for graphene fabrication can control the growth and nucleation of graphene and can generate graphene sheets of suitable size, shape and thickness, but have limitations of lending maximum amounts of impurities and defects to the structure. Cu foils used for the graphene synthesis depends upon the shape and size of Cu foil dimension. The nature of the metallic substrate is vital for the growth and nucleation of graphene. The first attempt to produce few-layer graphene films via the CVD method was reported using Cu foil and camphor as the carbon source. Metals with medium-high carbon solubility (>0.1 atomic %) such as Co and Ni support carbon diffusion at high temperatures [29]. The growth of graphene on substrates like Co and Ni is caused due to carbon diffusion (carbon solubility) on the metal thin film at the growth temperature to form high purity bulk materials after cooling of the surface [30]. Thermal CVD growth and deposition method is the most promising method for the fabrication of large-scale production of single or few-layer graphene films on various metal substrates. The growth of graphene using low carbon solubility substrates like Cu has showed impact on the fabrication and is considered a controlling factor for nucleation and growth on the surface. When compared to other metals Cu acts as the best catalyst and has many advantages such as cost, good controlling ability and ease of transfer. The main features of large-area synthesis on Cu depend upon (i) Cu to methane exposure and catalytic decomposition of methane on Cu from CxHy. Additionally, various parameters such as the growth temperature, methane pressure and flow and partial pressure of hydrogen are required for the synthesis process. (ii) It also depends on coverage and surface of Cu foil under the estimated temperature, and methane flow rate and pressure conditions can vary the production of graphene and (iii) nucleation or island formation of graphene (carbon source) depends upon the CxHy species, temperature and other flow rate conditions. Somani and co-workers first reported the synthesis of few-layer graphene films using the CVD method from camphor as the precursor on Ni foils [31]. This study has opened a new avenue for the synthesis of graphene but with few limitations such as controlling the number of layers and creating foldings or wrinkles in graphene. Thus, synthesis of graphene layers on several metal substrates with a controlled number of layers and foldings has been much focused [32–33]. The synthesis of graphene by the CVD method is additionally done by substitutional doping using nitrogen atoms doped on the surface of graphene. These nitrogen-doped graphene (N-graphene) layers have demonstrated interesting properties.

2.3 PLASMA ENHANCED CHEMICAL VAPOR DEPOSITION (PECVD)

Use of plasma is considered as one of the most promising methods to produce graphene with desired properties. After completion of the pyrolysis process, the system is cooled down naturally in inert atmosphere and residual black powder is collected. Hydrogen plasma treatment on residual black powder is carried out at 350°C for 3–4 min in a microwave plasma chemical vapor deposition apparatus operating at 2.45 GHz frequency. The effect of plasma on the structures of carbonaceous products was studied and it is revealed that the carbonaceous structures such as 3D nano GO, graphite nanodots, carbon nanotubes, and carbon onions were present in both pre-treated and plasma-treated residual black powder. For the synthesis of graphene PECVD is preferred over CVD due to its low reaction temperature, which minimizes cost of production. High amounts of graphite can be obtained using a DC discharge PECVD for the fabrication of "nanostructured graphite". The first production of mono- and few-layer graphene on several types of substrates by radio frequency PECVD was reported; it used a gas mixture of CH_4 and H_2 at 900 W and a reaction temperature of 680°C.

The preference of PECVD to CVD is due to its lower deposition time ($\leq$ 5 min) and lower growth temperature of 650°C. Also, PECVD has additional high-density reactive gas atoms and radicals, which allows low-temperature and rapid synthesis of graphene [22].

2.4 SYNTHESIS OF GRAPHENE SHEETS BY GASEOUS CARBON PRECURSORS

The advantages of using gaseous precursors for the synthesis of graphene or carbon-based materials are that gaseous precursors occupy less space than solid and liquid carbon precursors and additionally can very easily be stored in special tanks. The synthesis of graphene can be carried out using gas precursor beside the solids and liquid precursors. Generally, methane (CH_4), acetylene (C_2H_2) and ethylene (C_2H_4) are mostly used as carbon precursors for the growth and fabrication of graphene via the CVD method. Methane gas has been widely used as the carbon precursor for the growth of graphene on Cu catalyst although breaking of C-H bond in methane is very difficult and have poor reactivity. To produce high-quality graphene on a metallic substrate with methane as precursor requires a high temperature of above 1000°C [30]. Recently GuO et al. reported use of methane on molten Cu for developing monolayer graphene but molten Cu could not withstand longer time in the reaction chamber due to the rapid Cu evaporation in the growth stage in the chamber. Sun et al. fabricated large-area graphene on solid glass substrates using methane as precursor but with a considerable amount of defects in the graphene [34]. This procedure provides a cheaper route for large-scale production of graphene on solid glasses. Apart from methane, acetylene is widely used as carbon precursor for the growth of graphene via the CVD method. Chen et al. reported the growth of graphene films by reducing the flow rate of acetylene at 600°C in the LPCVD method. This reduction in the flow rate of acetylene played a vital role in the graphene synthesis, and monolayer graphene can be obtained on Cu metal in ultra-high vacuum CVD. The high rate of acetylene flow causes carbon smoke in the CVD chamber and hinders the growth and nucleation of graphene. Qi et al. developed bilayer graphene films in atmospheric conditions with acetylene on Cu foil and by varying the flow rate of H_2 and Ar gases [35–36]. The production of 3D networks of graphene on Ni foam can be achieved using ethylene at a temperature of 850°C and at favorable pressure conditions. Ethylene as carbon precursor showed much better nucleation and growth of graphene when compared to methane due to ethylene's high reactivity, which allows higher amounts of deposition on metal substrate. Similarly, Sagar et al. fabricated graphene using ethylene on Cu and Ni foils by varying the pressure of the reactor from 0.1 to 0.4 M Pa. In this study the flow rate and the pressure variation were deeply examined. The graphene synthesis via the CVD method using different gas precursors is summarized in Table 2.1.

2.5 SYNTHESIS OF GRAPHENE SHEETS BY LIQUID PRECURSOR

The fabrication of different graphene structures using liquid precursors has gained attention and interest of researchers. Srivastava et al. reported the synthesis of large area and uniform growth of graphene sheets using hexane in the CVD technique. This approach to fabrication is the stepping stone to the synthesis of large-area and defect-free uniform graphene using liquid precursors and in turn offers a facile route to dope graphene using various organic solvents as precursors [47]. Synthesis of mono-layer graphene has attracted researchers to increase or construct a bandgap of interested by doping different kinds of dopant. The development of monolayer graphene with a desired bandgap using different organic solvents makes the fabrication of graphene easy, cheap, and safe. Kishi et al. synthesized single-layer graphene using isopropyl alcohol on polycrystalline Ni substrate under the infrared lamp heating sources. This approach of graphene synthesis is not effective as it generates considerable defects in the structure, and changes in phases and the crystalline nature as confirmed by Raman spectroscopy. This experiment used ethanol instead of isopropyl alcohol, which gave a better single-layer graphene [48]. Miyata et al. used ethanol for the synthesis of monolayer graphene sheet via the flash cooling method just after the CVD process. This process of cooling provides a favorable growth condition for the nucleation and surface diffusion

TABLE 2.1 Graphene synthesis using gas precursors in the CVD method

CARBON PRECURSOR	*GROWTH SUBSTRATE*	*GROWTH CONDITIONS (TEMPERATURE=T, PRESSURE =P, AND FLOW RATE= F)*	*GRAPHENE MORPHOLOGY*	*REF*
Methane	Cu foil	T=1000 P= 580 m Torr; F: CH_4:H_2 = 5:50	Monolayer graphene	37
Ethylene C_2H_4	Ni foam Thickness = 1.8 mm, Size = 40 9 60 mm	T= 750–850; P= 1 atm; F= C_2H_4 15	3D graphene	38
Methane	Quartz glass, borosilicate glass, sapphire glass	T=1020; P=1 atm; F=Ar:CH_4:H_2 = 100:8:50	Monolayer graphene	39
Methane	Cu foil, Thickness = 25 micro m, Size = 10 × 10 mm.	T=650, P= 7 kPa; F= CH_4:H_2 = 5: 500	Monolayer graphene	40
Acetylene	Cu foil Thickness = 50 µm	T=100 C/min) 1035; P= 0.8 torr; F=C_2H_2:H_2 = 2.8:80	Monolayer graphene	41
Acetylene	Cu foil Thickness = 25 µm Area = 6 × 6 cm^2	T= 1000, P=1 atm; F= C_2H_2:H2: Ar = 1:100:900	Bilayer graphene	42
Acetylene	Cu and Ni foil Thickness = 25 µm Area = 1 × 1 cm^2	T=1050; P=0.2 MPa; F= C_2H_4:H2:Ar = 40:40:160	Mono and few-layer graphene	43
Acetylene	Cu film deposited on SiO_2/Si; Cu thickness = 1 µm	T= 860 C; P= 1 kPa; F= H_2:Ar = 100:1000	Bilayer graphene	44
Propene	Rh (111) foil Thickness = 3 mm	T= 700–1000 K; P= 2 X 10^{-8} mbar	Graphene with low-defect density	45
Acetylene	Ni foil Thickness = 25 µm	T=600; P= 133.3 Pa; F=C_2H_2:H_2 = 12:12	Few-layer graphene	46

of carbon atoms on the surface of the Ni substrate. Additionally, it restricts forming of bulk carbon precipitation. This cooling process is a better method for the synthesis of graphene using liquid precursors [49]. Li et al. reported the synthesis of single-layer graphene flakes using benzene as the liquid precursor at a very low reaction temperature of 300°C. The interconnected carbon atoms in benzene can be easily dehydrogenated and can be easily attached to the Cu substrate for better nucleation and growth of graphene. The graphene synthesis was not found on Cu substrate due to a very low temperature and insufficient energy at 200°C but laid a foundation that benzene can be used for graphene synthesis [50]. Guermoune reported the growth of single-layer graphene using methanol instead of ethanol and propanol on the Cu substrate under a pressure of 10^{-6} Torr. Use of methanol for graphene synthesis yielded better quality graphene structures, which were examined by Raman spectra [51]. Different liquid carbon precursors used for making graphene nanosheets are summarized in Table 2.2.

2.6 SYNTHESIS OF GRAPHENE SHEETS BY SOLID CARBON SOURCE

In the present trends, use of solid carbon sources in CVD method for the synthesis of graphene is very attractive because the growth of graphene from solid precursors is cost effective and non toxic. The use of solid precursors is advantageous because of easy handling, no toxic effect and less space requirement. Ruan et al. reported the synthesis of high-quality graphene from a carbon precursor at 1050°C under vacuum conditions. Materials rich in carbon such as grass, cookie, plastic, dog feces, chocolates and waste foods can be used as carbon precursors[61]. In the CVD method, solid carbon precursors are placed on top of Cu foil to produce monolayer graphene on the backside of Cu foil. Formation of high-quality graphene on the bottom side of Cu foil mainly depends upon the solubility of carbon at a high temperature of 1050 degree in the vacuum reaction chamber with a flow rate of Ar and H_2 at 600 sccm. Sun et al. reported the synthesis of N-doped monolayer graphene using PMMA at 800°C with melamine as nitrogen sources and of N-doped graphene with PMMA on the surface of Cu foil. Fluorene and sucrose were also used as carbon precursors to produce a high-quality graphene with small topological defects. The problem with sucrose for graphene synthesis is the presence of high concentration of heteroatoms, oxygen and fluorine includes a five-member ring structure [62]. The dissociation of C atoms at elevated temperatures results from the affinity of the Cu catalyst surface to O heteroatoms. Choi et

TABLE 2.2 Synthesis of graphene via CVD using liquid carbon precursors

CARBON PRECURSOR	*GROWTH SUBSTRATES*	*GROWTH CONDITIONS (TEMPERATURE, PRESSURE AND FLOW RATE)*	*GRAPHENE MORPHOLOGY*	*REF*
C_2H_5OH Ethanol	Ni Thickness = 25 μm, Size = 3 9 3 cm	T= 850; P= 1 Torr; F= H2 = 10	Monolayer graphene	52
Ethanol	Stainless steel- SS304 Thickness = 0.1 mm Size = 1 × 1 cm	T= 850; P= 1 Torr; F= H2 = 10	Mono-, bi-, tri-layer graphene	53
Pyridine, C_5H_5N	Cu	T= 300, P= 1 atm	Monolayer graphene and N-doped graphene	54
Methanol, CH_3OH Ethanol, 1-propanol, C2H5OH	Cu Thickness = 25 μm,	T= 850; P= 1 Torr; F= H_2 = 10	Monolayer graphene	55
Pentane, C_5H_{12}	Cu Thickness = 50 μm APCVD	T=900; P= 1 atm; F= Ar:H2 = 1000:10	Mono-, bi-, few-layer	56
Benzene, C_6H_6	Cu foil CVD	T=300; P= 8–15 Torr; H2 = 50	Monolayer grapheme flakes	57
C_6H_6	Cu Thickness = 25 μm Oxygen free APCVD	T=300; P= 1 atm; F= Ar:H2 = 5:20	Monolayer graphene	58
C_6H_6	Cu (111) Size = 12 9 5 mm2	T=437; P=1 × 10^{-10} mbar	Multilayer graphene	59
Cyclohexane, C_6H_{12} LPCVD	Cu2NiZn ternary alloy	T=100; P= 300–600 m Torr	Monolayer graphene	60

TABLE 2.3 Synthesis of graphene via CVD using solid carbon precursors

CARBON PRECURSOR	*GROWTH SUBSTRATES*	*GROWTH CONDITIONS (TEMPERATURE, PRESSURE AND FLOW RATE)*	*GRAPHENE MORPHOLOGY*	*REF*
Hexabenzocoronene $C_{48}H_{24}$	Co (0001) Thermal CVD	T=327; P= 3 × 10^{-11} Torr	Monolayer graphene	66
Asphalt	Ni foam CVD	T= 940; P= 1500 Pa; F= Ar:H_2 = 300:30	Multilayer 3D graphene	67
Amorphous carbon	Cu foil; Thickness = 25 μm; Size = 1X5 cm	T=1035; P=20 m Torr; F= H_2 = 2	Monolayer graphene	68
Highly oriented pyrolytic graphite (HOPG)	Ni deposited on HOPG Thickness = 100 nm Size = 2 ×2 cm	T=650; P= 5 ×10^{-8} Torr	Monolayer graphene	69
Camphor ($C_{10}H_{16}O$)	Cu foil APCVD	T=1020; P= 1 atm	Monolayer and bilayer graphene	70
Camphor	Ni foil Size = 5 ×5 mm APCVD	T= 870; P= 1 atm	Single-layer to few-layers graphene	71
Camphor	Ni foil Size = 2 × 2 cm2	T= 700–850; P= 1 atm	Few-layer graphene	72
Thiocamphor ($C_{10}H_{16}S$) APCVD	Cu foil, Thickness = 0.025 mm Size = 5 ×5 mm2	T=1000; P= 1 atm; F= Ar:H2 = 485:15	Few-layer thiolated graphene	73

al. synthesized monolayer graphene at 300°C with p-terphenyl as the carbon source. For the growth of graphene the high temperature can be reduced to a lower temperature by replacing methane with p-terphenyl [63]. Ray et al. reported the synthesis of nickel decorated with graphene nano powder using lotus and hibiscus flower petals at elevated temperature by thermal exfoliation. Use of Ni enhances the graphene electron density near the Fermi energy level. Using the flower petals as carbon source can avoid production of toxic and hazardous substances during the large-scale synthesis of graphene [64]. This method is an easy and eco-friendly for graphene synthesis. High temperatures are needed to facilitate the removal of oxygen-containing molecules. Recently cheaper natural carbon sources are used on metal substrates. Another interesting graphene source as a carbon precursor is asphalt, which contains large amounts of carbon compounds with alkyl chains linked to aromatic cores [65]. Different solid carbon precursors used for making graphene nanosheets are summarized in Table 2.3.

CONCLUSION

In conclusion, this review gave insightful details and summary of graphene synthesis from different types of carbon precursors using the CVD method. In particular the use of solid, liquid and gaseous carbon precursors for the synthesis of graphene nanosheets was discussed. The important roles of carbon precursors and the optimal conditions for the growth and nucleation of graphene nanosheets are also summarized. In-depth study on graphene synthesis and its potential applications are discussed in detail in the next chapter. Fabrication of large and scalable graphene from camphor is a novel idea and further research on different carbon sources is needed. A very simple and cost-effective method is developed to synthesize single- and multi-layer graphene sheets using camphor as a solid carbon source. A controlled island formation for single and multi-layered graphene sheets was fabricated using camphor source. A defect-free and controlled bonding of carbon led to the formation of ordered graphene layers.

REFERENCES

1. K. S. Novoselov, A. K. Geim, S. V. Morozov, D. Jiang, M. I. Katsnelson, I. V. Grigorieva, S. V. Dubonos, and A. A. Firsov, "Two-dimensional gas of massless Dirac fermions in graphene," *Nature*, vol. 438, p. 197, 2005.
2. A. K. Geim, and K. S. Novoselov, "The rise of graphene," *Nat. Mater.*, vol. 6, p. 183, 2007.
3. K. S. Novoselov, A. K. Geim, S. V. Morozov, D. Jiang, Y. Zhang, S. V. Dubonos, I. V. Grigorieva, and A. A. Firsov, "Electric field effect in atomically thin carbon films," *Science*, vol. 306, no. 5696, pp. 666–669, 2004.
4. A. K. Geim, "Graphene: status and prospects," *Science*, vol. 324, no. 5934, pp. 1530–1534, 2009.

5. L. Peng, Z. Xu, Z. Liu, Y. Guo, P. Li, and C. Gao, "Ultrahigh thermal conductive yet superflexible graphene films," *Adv. Mater.*, vol. 29, p. 1700589, 2017.
6. K. S. Novoselov, Z. Jiang, Y. Zhang, S. V. Morozov, H. L. Stormer, U. Zeitler, et al., "Room-temperature quantum hall effect in grapheme," *Science*, vol. 315, no. 5817, p. 1379, 2007.
7. K. I. Bolotin, K. J. Sikes, Z. Jiang, M. Klima, G. Fudenberg, J. Hone, et al., "Ultrahigh electron mobility in suspended grapheme," *Solid State Commun.*, vol. 146, no. 9–10, pp. 351–355, 2008.
8. C. G. Lee, X. D. Wei, J. W. Kysar, and J. Hone, "Measurement of the elastic properties and intrinsic strength of monolayer grapheme," *Science*, vol. 321, no. 5887, pp. 385–388, 2008.
9. A. A. Balandin, S. Ghosh, W. Bao, I. Calizo, D. Teweldebrhan, F. Miao, et al., "Superior thermal conductivity of single-layer graphene," *Nano Lett.*, vol. 8, no. 3, pp. 902–907, 2008.
10. M. Orlita, C. Faugeras, P. Plochocka, P. Neugebauer, G. Martinez, D. K. Maude, et al., "Approaching the dirac point in high mobility multilayer epitaxial grapheme," *Phys. Rev. Lett.*, vol. 101, no. 26, p. 267601, 2008.
11. D. R. Dreyer, S. Park, C. W. Bielawski, and R. S. Ruoff, "The chemistry of graphene oxide," *Chem. Soc. Rev.*, vol. 39, no. 1, pp. 228–240, 2010.
12. S. Park, and R. S. Ruoff, "Chemical methods to produce graphenes," *Nature Nanotechnol.*, vol. 29, no. 4, pp. 217–224, 2009.
13. F. S. Kim, Y. Zhao, H. Jang, S. Y. Lee, J. M. Kim, K. S. Kim, et al., "Largescale pattern growth of graphene films for stretchable transparent electrodes," *Nature*, vol. 457, no. 7230, pp. 706–710, 2009.
14. X. Lu, M. Yu, H. Huang, and R. S. Ruoff, "Tailoring graphite with the goal of achieving single sheets," *Nanotechnology*, vol. 10, no. 3, pp. 269–272, 1999.
15. C. Berger, Z. Song, X. Li, X. Wu, N. Brown, C. Naud, et al., "Electronic confinement and coherence in patterned epitaxial grapheme," *Science*, vol. 312, no. 5777, pp. 1191–1196, 2006.
16. M. Choucair, P. Thordarson, and J. A. Stride, "Gram-scale production of graphene based on solvothermal synthesis and sonication," *Nature Nanotechnol.*, vol. 4, no. 1, pp. 30–33, 2009.
17. S. Stankovich, "Synthesis of graphene-based nanosheets via chemical reduction of exfoliated graphite oxide," *Carbon*, vol. 45, no. 3, pp. 1558–1565, 2007.
18. W. S. Hummers, and R. E. Offman, "Preparation of graphene oxide," *J. Am. Chem. Soc.*, vol. 80, p. 1339, 1958.
19. L. Staudenmaier, "Process for the preparation of graphitic acid," *Ber. Dtsch. Chem. Ges.*, vol. 31, p. 1481, 1898.
20. B. C. Brodie, "On the atomic weight of graphite," *Philos. Trans. R. Soc. Lond.*, vol. 149, p. 249, 1859.
21. S. Park, and R. Ruoff, "Chemical methods for the production of graphenes," *Nat. Nanotechnol.*, vol. 4, pp. 217–224, 2009.
22. A. J. Van Bommel, J. E. Crombeen, and A. Van Tooren, "LEED and Auger electron observations of the SiC(0001)," *Surf. Sci.*, vol. 48, no. 2, pp. 463–472, 1975.
23. D. A. C. Brownson, D. K. Kampouris, and C. E. Banks, "Graphene electrochemistry: fundamental concepts through to prominent applications," *Chem. Soc. Rev.*, vol. 41, pp. 6944–6976, 2012.
24. G. H. Lee, R. C. Cooper, S. J. An, S. Lee, A. v. d. Zande, N. Petrone, Al. G. Hammerberg, C. Lee, B. Crawford, W. Oliver, J. W. Kysar, and J. Hone,

"High-strength chemical-vapor–deposited graphene and grain boundaries," *Sciences*, vol. 340, pp. 1073–1076, 2013.
25. J. Jang, M. Son, S. Chung, et al., "Low-temperaturegrown continuous graphene films from benzene by chemical vapor deposition at ambient pressure," *Sci. Rep.*, vol. 5, p. 17955, 2015.
26. K. Lee, and J. Ye, "Significantly improved thickness uniformity of graphene monolayers grown by chemical vapor deposition by texture and morphology control of the copper foil substrate," *Carbon*, vol. 100, pp. 441–449, 2016.
27. K. Choy, "Chemical vapour deposition of coatings," *Prog. Mater. Sci.*, vol. 48, pp. 57–170, 2003.
28. W.-W. Liu, S.-P. Chai, A. R. Mohamed, and U. Hashim, "Synthesis and characterization of graphene and carbon nanotubes: a review on the past and recent developments," *J. Ind. Eng. Chem.*, vol. 20, no. 4, pp. 1171–1185, 2014.
29. A. Reina, X. Jia, J. Ho, D. Nezich, H. Son, V. Bulovic, et al., "Large area, few-layer graphene films on arbitrary substrates by chemical vapor deposition," *Nano Lett.*, vol. 9, pp. 30, 2008.
30. P. R. Somani, S. P. Somani, and M. Umeno, "Planer nano-graphenes from camphor by CVD," *Chem Phys Lett.*, vol. 430, p. 56, 2006.
31. K. S. Kim, Y. Zhao, H. Jang, S. Y. Lee, J. M. Kim, K. S. Kim, et al., "Large-scale pattern growth of graphene films for stretchable transparent electrodes," *Nature*, vol. 457, p. 706, 2009.
32. X. Li, W. Cai, J. An, S. Kim, J. Nah, D. Yang, et al., "Large-area synthesis of high-quality and uniform graphene films on copper foils," *Science*, vol. 324, p. 1312, 2009.
33. Y. Lee, S. Bae, H. Jang, S. Jang, S.-E. Zhu, S. H. Sim, et al., "Wafer-scale synthesis and transfer of graphene films," *Nano Lett.*, vol. 10, p. 409, 2010.
34. J. Sun, Y. Chen, M. K. Priydarshi, et al., "Direct chemical vapor deposition-derived graphene glasses targeting wide ranged applications," *Nano Lett.*, vol. 15, pp. 5846–5854, 2015.
35. N. S. Mueller, A. J. Morfa, and D. Abou-Ras, et al., "Growing graphene on polycrystalline copper foils by ultra-high vacuum chemical vapor deposition," *Carbon*, vol. 78, pp. 347–355, 2014.
36. M. Qi, Z. Ren, Y. Jiao, et al., "Hydrogen kinetics on scalable graphene growth by atmospheric pressure chemical vapor deposition with acetylene," *J. Phys. Chem. C.*, vol. 117, pp. 14348–14353, 2013.
37. K. Lee, and J. Ye, "Significantly improved thickness uniformity of graphene monolayers grown by chemical vapor deposition by texture and morphology control of the copper foil substrate," *Carbon*, vol. 100, pp. 441–449, 2016.
38. P. Trinsoutrot, H. Vergnes, and B. Caussat, "Three-dimensional graphene synthesis on nickel foam by chemical vapor deposition from ethylene," *Mater. Sci. Eng. B Solid State Mater. Adv. Technol.*, vol. 179, pp. 12–16, 2014.
39. J. Sun, Y. Chen, M. K. Priydarshi, et al., "Direct chemical vapor deposition-derived graphene glasses targeting wide ranged applications," *Nano Lett.*, vol. 15, pp. 5846–5854, 2015.
40. L. Fang, W. Yuan, B. Wang, and Y. Xiong, "Growth of graphene on Cu foils by microwave plasma chemical vapour deposition: the effect of in situ hydrogen plasma posttreatment," *Appl. Surf. Sci.*, vol. 383, pp. 28–32, 2016.

41. M. Yang, S. Sasaki, K. Suzuki, H. Miura, "Control of the nucleation and quality of graphene grown by low pressure chemical vapor deposition with acetylene," *Appl. Surf. Sci.*, vol. 366, pp. 219–226, 2016.
42. M. Qi, Z. Ren, Y. Jiao, et al., "Hydrogen kinetics on scalable graphene growth by atmospheric pressure chemical vapor deposition with acetylene," *J. Phys. Chem. C.*, vol. 117, pp. 14348–14353, 2013.
43. R. R. Sagar, X. Zhang, and C. Xiong, "Growth of graphene on copper and nickel foils via chemical vapour deposition using ethylene," *Mater. Res. Innovat.*, vol. 18, pp. S4706–S4710, 2014.
44. K. Yagi, A. Yamada, K. Hayashi, et al., "Dependence of field-effect mobility of graphene grown by thermal chemical vapor deposition on its grain size," *Jpn. J. Appl. Phys.*, vol. 52, p. 110106, 2013.
45. K. Gotterbarm, W. Zhao, and O. Höfert, et al., "Growth and oxidation of graphene on Rh(111)," *Phys. Chem. Chem. Phys.*, vol. 15, p. 19625, 2013.
46. C. S. Chen, and C. K. Hsieh, "Effects of acetylene flow rate and processing temperature on graphene films grown by thermal chemical vapor deposition," *Thin Solid Films*, vol. 584, pp. 265–269, 2015.
47. A. Srivastava, C. Galande, and L. Ci, "Novel liquid precursor-based facile synthesis of large-area continuous, single, and few-layer graphene films," *Chem. Mater.*, vol. 22, pp. 3457–3461, 2010.
48. N. Kishi, A. Fukaya, R. Sugita, et al., "Synthesis of graphenes on Ni foils by chemical vapor deposition of alcohol with IR-lamp heating," *Mater. Lett.*, vol. 79, pp. 21–24, 2012.
49. Y. Miyata, K. Kamon, and K. Ohashi, "A simple alcoholchemical vapor deposition synthesis of single-layer graphenes using flash cooling," *Appl. Phys. Lett.*, vol. 96, p. 263105, 2010.
50. Z. Li, P. Wu, C. Wang, et al., "Low-temperature growth of graphene by chemical vapor deposition using solid and liquid carbon sources," *ACS Nano.*, vol. 5, pp. 3385–3390, 2011.
51. A. Guermoune, T. Chari, F. Popescu, et al., "Chemical vapor deposition synthesis of graphene on copper with methanol, ethanol, and propanol precursors," *Carbon*, vol. 49, pp. 4204–4210, 2011.
52. Y. Miyata, K. Kamon, and K. Ohashi, "A simple alcoholchemical vapor deposition synthesis of single-layer graphenes using flash cooling," *Appl. Phys. Lett.*, 96, p. 263105, 2010.
53. R. John, A. Ashokreddy, C. Vijayan, and T. Pradeep, "Single- and few-layer graphene growth on stainless steel substrates by direct thermal chemical vapor deposition," *Nanotechnology*, vol. 22, p. 165701, 2011.
54. Y. Xue, B. Wu, L. Jiang, et al., "Low temperature growth of highly nitrogen doped single graphene arrays by chemical vapor deposition," *J. Am. Chem. Soc.*, vol. 134, pp. 11060–11063, 2012.
55. A. Guermoune, T. Chari, F. Popescu, et al., "Chemical vapor deposition synthesis of graphene on copper with methanol, ethanol, and propanol precursors," *Carbon*, vol. 49, pp. 4204–4210, 2011.
56. X. Dong, P. Wang, W. Fang, et al., "Growth of large sized graphene thin-films by liquid precursor-based chemical vapor deposition under atmospheric pressure," *Carbon*, vol. 49, pp. 3672–3678, 2011

57. J. -H. Choi, Z. Li, P. Cui, et al., "Drastic reduction in the growth temperature of graphene on copper via enhanced London dispersion force," *Sci. Rep.*, vol. 3, p. 1925, 2013.
58. J. Chang, M. Son, S. Chung, et al., "Low-temperature grown continuous graphene films from benzene by chemical vapor deposition at ambient pressure," *Sci. Rep.*, vol. 5, p. 17955, 2015.
59. Q. Han, H. Shan, J. Deng, et al., "Construction of carbon-based two-dimensional crystalline nanostructure by chemical vapor deposition of benzene on Cu(111)," *Nanoscale*, vol. 6, pp. 7934–7939, 2014.
60. W. Gan, N. Han, C. Yang, et al., "A ternary alloy substrate to synthesize monolayer graphene with liquid carbon precursor," *ACS Nano*, vol. 11, pp. 1371–1379, 2017.
61. M. J. Salifairus, S. B. Abd Hamid, T. Soga, et al., "Structural and optical properties of graphene from green carbon source via thermal chemical vapor deposition," *J. Mater. Res.*, vol. 31, pp. 1–10, 2016.
62. Z. Z. Sun, Z. Yan, J. Yao, et al., "Growth of graphene from solid carbon sources," *Nature*, vol. 468, pp. 549–552, 2010.
63. J.-H. Choi, Z. Li, P. Cui, et al., "Drastic reduction in the growth temperature of graphene on copper via enhanced London dispersion force," *Sci. Rep.*, vol. 3, p. 1925, 2013.
64. A. K. Ray, R. K. Sahu, V. Rajinikanth, et al., "Preparation and characterization of graphene and Ni-decorated graphene using flower petals as the precursor material," *Carbon*, vol. 50, pp. 4123–4129, 2012.
65. I. F. Cheng, Y. Xie, R. Allen Gonzales, et al., "Synthesis of graphene paper from pyrolyzed asphalt," *Carbon*, vol. 49, pp. 2852–2861, 2011.
66. D. Eom, D. Prezzi, K. T. Rim, et al., "Structure and electronic properties of graphene nanoislands on Co(0001)," *Nano Lett.*, vol. 9, pp. 2844–2848, 2009.
67. Z. Liu, Z. Tu, Y. Li, et al., "Synthesis of three-dimensional graphene from petroleum asphalt by chemical vapor deposition," *Mater. Lett.*, vol. 122, pp. 285–288, 2014.
68. H. Ji, Y. Hao, Y. Ren, et al., "Graphene growth using a solid carbon feedstock and hydrogen," *ACS Nano.*, vol. 5, pp. 7656–7661, 2011.
69. M. Xu, D. Fujita, K. Sagisaka, et al., "Production of extended single-layer," *ACS Nano.*, vol. 5, pp. 1522–1528, 2011.
70. G. Kalita, K. Wakita, and M. Umeno, "Monolayer graphene from a green solid precursor," *Physica. E*, vol. 43, pp. 1490–1493, 2011.
71. M. Ahmed, N. Kishi, R. Sugita, et al., "Graphene synthesis by thermal chemical vapor deposition using solid precursor," *J. Mater. Sci. Mater. Electron.*, vol. 24, pp. 2151–2155, 2013.
72. H. Liu, N. Kishi, and T. Soga, "Synthesis of thiolated few layered graphene by thermal chemical vapor deposition using solid precursor," *Mater. Lett.*, vol. 159, pp. 114–117, 2015.
73. J. M. González-Domínguez, A. Colusso, L. Litti, A. Ostric, M. Meneghetti, T. Da Ros, "Thiolated graphene oxide nanoribbons as templates for anchoring gold nanoparticles: two-dimensional nanostructures for SERS," *ChemPlusChem*, vol. 84, p. 862, 2019.

Synthesis of graphene by natural camphor

3

Harsh A. Chaliyawala and
Indrajit Mukhopadhyay

Contents

3.1 INTRODUCTION

Since its invention in 2004, graphene has attracted much interest owing to its extraordinary properties such as quantum transport, optical transmittance, superior mobility, thermal conductivity, and superior mechanical properties, as discussed in the early chapters. A honeycomb lattice arrangement consisting of a long chain of carbon atoms in sp^2 hybridization offers numerous optoelectronic applications. Such high accuracy and large-scale optical devices demand graphene film with very high quality and large domain size on a larger area [1]. There are various processes currently under development to achieve MLG sheets over a larger area to attain a low sheet resistance and high transmittance values [2], [3]. Among them, the CVD process is one of the most versatile techniques to synthesize large-area graphene with a controllable number of layers and single-crystal domains. In this technique, the carbon precursors are decomposed on the Cu surface, for instance, due to their low carbon solubility and the high self-limiting factor [2], [4]–[7]. The resulting graphene on a metal substrate can be easily transferred to any desirable substrate by a simple polymer transferring process without affecting much of the properties of graphene.

Recently, gaseous hydrocarbons and liquid carbon sources have been replaced by forthcoming solid carbon sources such as camphor ($C_{10}H_{16}O$), polystyrene, etc., by a very facile and low-cost CVD approach to achieve a large area MLG on Cu foil with an optimum optoelectronic property [2], [5], [6]. The growth of mono/bi-layer graphene by employing camphor as a carbon source on different metal substrates has been studied in recent years as it promises to grow single crystals of graphene domains [8]–[10]. In recent years, significant efforts have been made to synthesize single-crystal MLG sheets from camphor on Cu foil to form a large area by the APCVD process. However, various factors such as gas composition, growth temperature, the flow rate of carbon source, and substrate surface purity have affected the graphene nucleation and growth on the Cu surface via APCVD[10]–[13]. There are various reports on the variation of annealing and gaseous flow rates to control the nucleation and growth kinetics of graphene crystals on the substrate surface [14]–[16]. Besides, various temperature conditions indicate the controlled formation of a monolayer, bi-layer, hexagonal structure, layer-stacking, and dendrite growth of graphene [17], [18]. In this regard, our approach is to investigate MLG on Cu foil with a large domain size that can deliver good optical and electrical properties. An efficient, low cost and facile approach using camphor as a carbon precursor will be utilized to develop an MLG sheet on Cu foil and transfer it on glass, Si/SiO_2 and flexible PET substrates.

3.2 ISLAND FORMATION OF MONO/ BI-LAYER GRAPHENE (MLG) SHEETS

The synthesis and formation of large-area graphene sheets (~1 cm × 1.5 cm) were performed using camphor, a botanical hydrocarbon source obtained by pyrolysis in a high-temperature atmospheric chemical vapour deposition (CVD) system. Camphor is a natural solid, low cost and environmentally friendly hydrocarbon. It consists of hexagonal and pentagonal rings along with methyl carbon structure, which plays an important role in graphene synthesis.

During pyrolysis, methyl carbon can be easily detached and fused to a hexagonal carbon ring to form a larger graphene sheet on Cu foil. During the growth process, at a certain distance (D) and an evaporation temperature (T_e) of ~350–400°C, the transformation of solid camphor to camphor vapor can take place, and under the Ar:H_2 flow the camphor molecules get transported to Cu foil. It is to be noted that the graphene nucleation from camphor vapor on Cu foil is a surface reaction as demonstrated for the synthesis of methane-based graphene by Li et al. [13]. Hereafter, catalytic decomposition of camphor to carbon atoms on Cu foil takes place at a very high temperature of 1020°C. Optimizing camphor concentration, distance between source and substrates, flow rates and growth time gives a continuous island formation of single/bi-layer carbon atoms on a larger surface that is ideally called a mono/bi-layer graphene sheet (MLG) as schematically shown in Figure 3.1 [19]. The as-grown graphene sheets can be easily transferred to any desired substrates by employing a polymer transfer process. In this process, graphene covered on Cu foil was coated with polymethyl methacrylate (PMMA) by spin coating at 3000 rpm for 60 seconds and 1000 rpm for 60 seconds, followed by baking of PMMA/graphene/copper at 180°C for 2 minutes as schematically shown in Figure 3.2. Then the copper was etched by placing the samples over 0.5 M $(NH_4)_2S_2O_8$ (Ammonium persulfate) for a few hours. The graphene grown on the backside of the Cu was removed by washing under the stream of DI water and placed again on the etching solution. The PMMA/graphene was repeatedly cleaned in DI water five to six times. Then the PMMA/graphene was transferred onto the

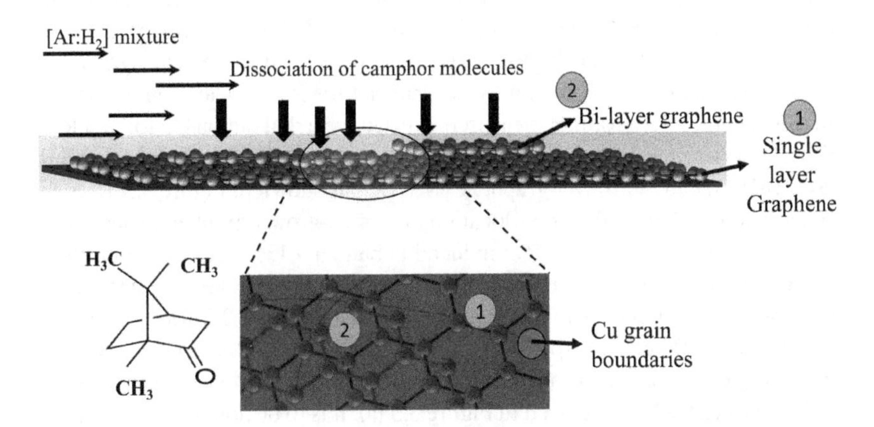

FIGURE 3.1 Represents the dissociation of camphor molecules at a high temperature (T= 1020°C) under Ar:H2 mixture by using the atmospheric chemical vapor deposition (APCVD) technique. The magnified image represents the formation of a single and bi-layer graphene sheet on Cu foil during the process along with the formation of Cu grain boundaries after annealing treatment.

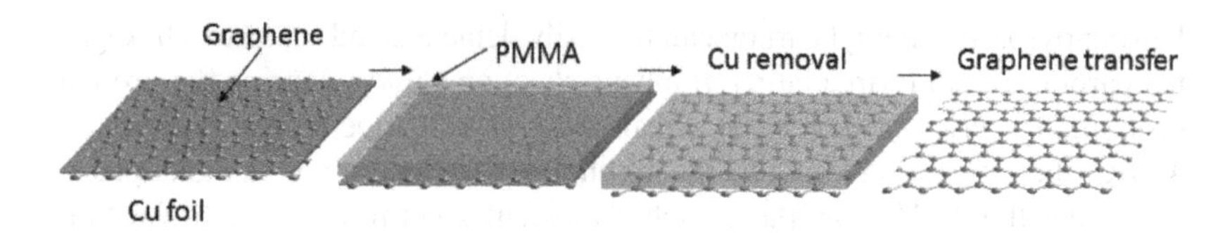

FIGURE 3.2 Removal of graphene coated with Cu foil by using a polymer transfer process.

glass, flexible PET and SiO_2/Si substrates. The substrates were kept overnight to facilitate complete adhesion. The PMMA layer was removed by dipping the PMMA coated graphene substrates in acetone for 30 minutes and then repeatedly washing with DI water and IPA. Finally, the graphene sheet can be easily transferred through a suitable method to any desired substrates such as Si/SiO_2 substrates, flexible PET films and soda lime glass substrates.

3.3 CHARACTERIZATION AND IDENTIFICATION OF GRAPHENE

The continuous growth of MLG sheets on a larger scale depends on the Cu surface purity, gas flow rates, growth temperature, camphor concentration and deposition time. Therefore, the optimization of all the physical parameters challenged us to develop high-quality, defect-free and large-scale graphene sheets by using natural camphor as a carbon source for optoelectronic devices. In this regard, micro Raman measurements were carried out on Cu foil to identify the crystalline nature, the number of layers and defects states of mono/bilayer (MLG) and multilayer graphene (MULG) as shown in Figure 3.3. Raman spectra were taken at different locations in an as-grown graphene sheet. The spectra showed almost no defect-induced D band at ~1360 cm^{-1} at the particular spot, representing a high quality of graphene crystal. Moreover, there is the presence of a characteristic Graphitic G and second order 2D Raman peak at ~1590 and ~2700 cm^{-1}, respectively. It has been observed that 2D band due to two phonon resonance is much more intense than that of the G band, which indicates an MLG as displayed in Figure 3.3 (a). It is to be noted that, at most of the places we found single-layer graphene but, at some places, a possibility of bi-layer graphene has been confirmed due to the island formation of graphene crystals. The 2D/G ratio for the MLG sheet is found to be 2.3, which suggests a single layer of graphene sheet with the corresponding FWHM of 2D peak 24 cm^{-1}. On the other hand, if the concentration of camphor is increased above a

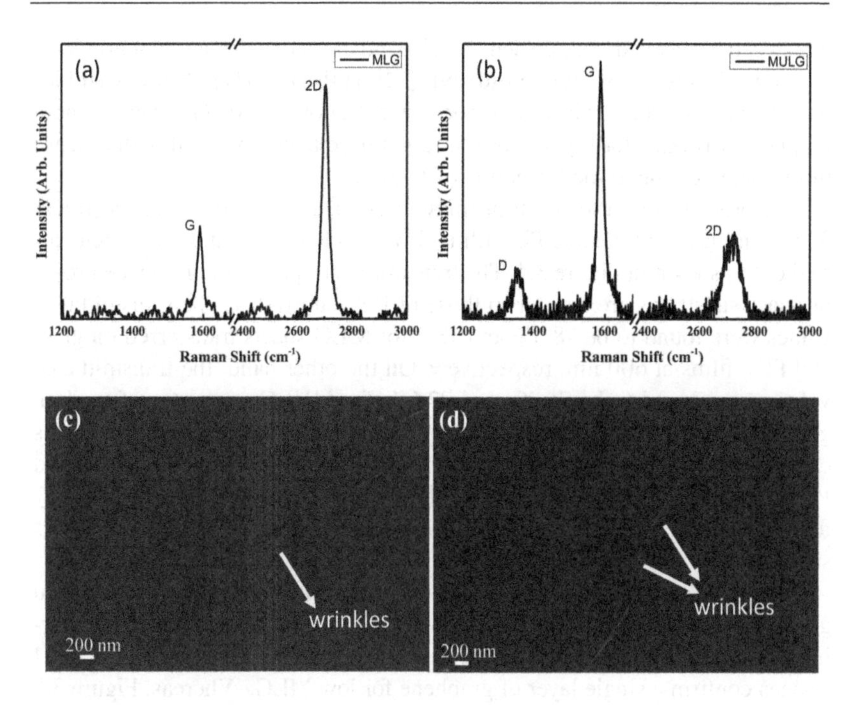

FIGURE 3.3 Micro Raman measurements were carried out on Cu foil to determine the (a) mono/bi-layer and (b) multi-layer graphene sheets. The surface morphological images of graphene for (c) 3.5 mg and (d) 5 mg camphor concentrations.

certain limit, there is an increase in the number of nucleation sites of graphene crystals which can form a stack of graphene layers. The 2D peak position can experience a blue shift with the increasing number of graphene layers, providing a signature of multi-layer graphene (MULG) as shown in Figure 3.3 (b) [12], [17]. The intensity was found to be reduced with a significant broadening of the 2D band, which suggests the presence of an MULG sheet. This effect introduced more defect states described as D band at ~1360 cm^{-1}. Therefore, a trade-off between the camphor concentration and the source to substrate distance will provide us the desired number of graphene layers. Furthermore, FESEM images of graphene sheets grown for 3.5 and 5 mg concentrations of camphor kept at constant distances, respectively, are shown in Figure 3.3 (c) and (d). The surface morphological images for the above two cases have been determined by transferring the graphene sheet on Si/SiO_2 substrates by a polymer transfer process as discussed in Chapter 1 [20]. Interestingly, graphene sheets have been homogenously covered over a large area for the abovementioned parameters. Moreover, there is a formation of graphene wrinkles, which can be easily distinguished and also indicates the formation of graphene as shown in Figure 3.3 (c) and (d) [9], [12], [14], [15], [17]. A few wrinkles have been observed for low camphor concentrations of MLG, whereas more graphene wrinkles for higher concentration have been observed with a dense morphology owing to the formation of MULG.

In order to observe the transmission characteristics of graphene transferred on glass and flexible PET films, UV-Vis characterization has been carried out as shown in Figure 3.4. The transmittance spectra for graphene grown on glass substrates are similar to those of PET substrates. The transmittance values were found to be 78.5% and 78% for MLG sheets transferred on glass and PET films at 600 nm, respectively. On the other hand, the transmittance values were found to be 75.0% and 72.6%, for MULG sheets transferred on glass and PET films, respectively. The results thus indicate that MLG sheets can give better transparent sheets than MULG sheets transferred on glass and PET films due to more carbon layers. The sheet resistance was nearly 800 Ω/□ for MLG and 1.5–2 KΩ/□ for MULG sheets. Higher sheet resistance offers successful optoelectronic applications.

HRTEM images of as-transferred graphene sheet on a carbon-coated Cu grid are shown in Figure 3.5. The edge of the graphene film folds back, allowing us to identify the cross-sectional view of the graphene sheet. The HRTEM images confirm a single layer of graphene for low MLG. Whereas, Figure 3.5 (c) indicates that a high concentration of camphor will lead us to more than ten layers of the graphene sheet. From the observations, the inter-planar spacing of

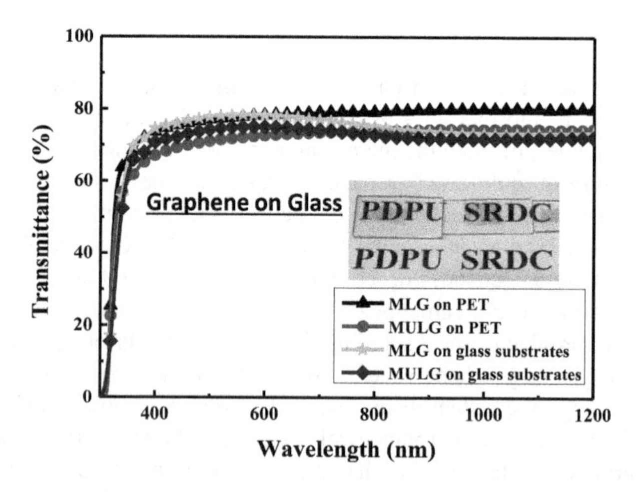

FIGURE 3.4 Represents transmittance spectra of a graphene sheet transferred on glass substrates. Inset shows an optical image of transferred MLG sheets on glass substrates.

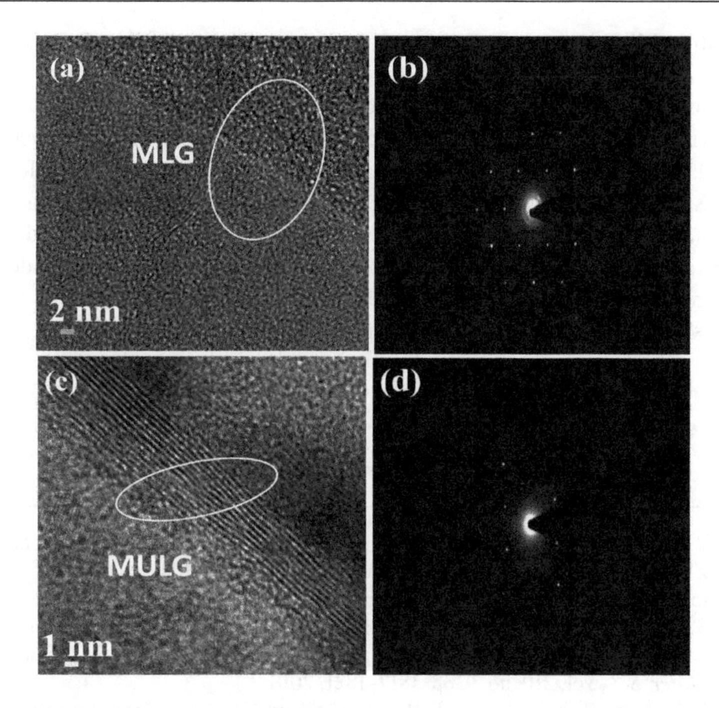

FIGURE 3.5 High-resolution transmission electron microscopic images of MLG and MULG sheets. The SAED patterns of MLG and MULG demonstrate the crystalline nature of graphene sheets.

0.38 nm for MULG was estimated to be exactly similar to that described in the existing literature [9], [17], [21]. Furthermore, we have carried out a complete survey by obtaining SAED patterns as shown in Figure 3.5 (b) and (d), respectively. The corresponding SAED pattern shows sharp and clear spots with a single crystal characteristic confirming a hexagonal structure for an MLG sheet. Alternatively, the intensity of the diffraction spot decreases, confirming the existence of polycrystalline nature, due to increases in the number of layers in multilayer graphene. The corresponding inter-planar d spacing value of 0.21 nm has been estimated for MLG as well as MULG sheet [1], [22].

CONCLUSION

In conclusion, we have presented a study to explain the effect of camphor concentration along with the source position for achieving a homogenous island growth of MLG sheets using natural camphor by the very facile APCVD technique. The Raman study confirms the existence of intense 2D band at ~2700 cm^{-1} and G band at 1590 cm^{-1} for the slow growth process at the rate of ~20 mg/min with 3.5 mg camphor kept at 16 cm. The existence of mono-/bilayer and multi-layer formation of homogeneous graphene sheets is confirmed by HRTEM images. Synthesized graphene sheets at various camphor concentrations at typical distances can be transferred onto glass and flexible PET substrates to identify the optimum electrical and optical properties. The measurement shows a decent sheet resistance value of ~1 kΩ/sq with a transmittance value of 78.3 % for MLG sheets transferred on glass and PET films at 600 nm.

REFERENCES

1. Z. Lin, T. Huang, X. Ye, M. Zhong, L. Li, J. Jiang, W. Zhang, L. Fan, and H. Zhu, "Thinning of large-area graphene film from multilayer to bilayer with a low-power CO_2 laser," *Nanotechnology*, vol. 24, p. 275302, 2013.
2. S. K. Behura, I. Mukhopadhyay, A. Hirose, Q. Yang, and O. Jani, "Vertically oriented few-layer graphene as an electron field-emitter," *Phys. Status Solidi Appl. Mater. Sci.*, vol. 210, no. 9, pp. 1817–1821, 2013.
3. P. Nguyen, S. K. Behura, M. R. Seacrist, and V. Berry, "Intergrain diffusion of carbon radical for wafer-scale, direct growth of graphene on silicon-based dielectrics," *ACS Appl. Mater. Interfaces*, vol. 10, no. 31, pp. 26517–26525, 2018.
4. J. Kraus, L. Böbel, G. Zwaschka, and S. Günther, "Understanding the reaction kinetics to optimize graphene growth on Cu by chemical vapor deposition," *Ann. Phys.*, vol. 529, no. 11, pp. 1–16, 2017.
5. X. Chen, L. Zhang, and S. Chen, "Large area CVD growth of graphene," *Synth. Met.*, vol. 210, no. December, pp. 95–108, 2015.
6. X. Zhang, L. Wang, J. Xin, B. I. Yakobson, and F. Ding, "Role of hydrogen in graphene chemical vapor deposition growth on a copper surface," *J. Am. Chem. Soc.*, vol. 136, no. 8, pp. 3040–3047, 2014.
7. Z. Sun, Z. Yan, J. Yao, E. Beitler, Y. Zhu, and J. M. Tour, "Growth of graphene from solid carbon sources," *Nature*, vol. 468, no. 7323, pp. 549–552, 2010.
8. G. Kalita, R. Hirano, M. E. Ayhan, and M. Tanemura, "Fabrication of a Schottky junction diode with direct growth graphene on silicon by a solid phase reaction," *J. Phys. D. Appl. Phys.*, vol. 46, no. 45, p. 455103, 2013.
9. G. Kalita, K. Wakita, and M. Umeno, "Monolayer graphene from a green solid precursor," *Phys. E Low-Dimensional Syst. Nanostructures*, vol. 43, no. 8, pp. 1490–1493, 2011.
10. M. T. G. Kalita, T. Sugiura, Y. Wakamatsu, and R. Hirano, "Controlling direct growth of graphene on insulating substrate by solid phase reaction of polymer layer," *RCS Adv.*, vol. 4, pp. 38450–38454, 2014.

11. G. Kalita, K. Wakita, M. Takahashi, and M. Umeno, "Iodine doping in solid precursor-based CVD growth graphene film," *J. Mater. Chem.*, vol. 21, no. 39, pp. 15209–15213, 2011.
12. A. Ibrahim, S. Akhtar, M. Atieh, R. Karnik, and T. Laoui, "Effects of annealing on copper substrate surface morphology and graphene growth by chemical vapor deposition," *Carbon N. Y.*, vol. 94, pp. 369–377, 2015.
13. P. Li, Z. Li, and J. Yang, "Dominant kinetic pathways of graphene growth in chemical vapor deposition: the role of hydrogen," *J. Phys. Chem. C*, vol. 121, no. 46, pp. 25949–25955, 2017.
14. D. Ding, P. Solís-Fernández, R. M. Yunus, H. Hibino, and H. Ago, "Behavior and role of superficial oxygen in Cu for the growth of large single-crystalline graphene," *Appl. Surf. Sci.*, vol. 408, pp. 142–149, 2017.
15. R. Papon, G. Kalita, S. Sharma, S. M. Shinde, R. Vishwakarma, and M. Tanemura, "Controlling single and few-layer graphene crystals growth in a solid carbon source based chemical vapor deposition," *Appl. Phys. Lett.*, vol. 105, no. 13, 2014.
16. S. Sharma, G. Kalita, R. Hirano, Y. Hayashi, and M. Tanemura, "Influence of gas composition on the formation of graphene domain synthesized from camphor," *Mater. Lett.*, vol. 93, no. February 2013, pp. 258–262, 2013.
17. K. M. Subhedar, I. Sharma, and S. R. Dhakate, "Control of layer stacking in CVD graphene under quasi-static condition," *Phys. Chem. Chem. Phys.*, vol. 17, no. 34, pp. 22304–22310, 2015.
18. T. Wu, et al., "Continuous graphene films synthesized at low temperatures by introducing coronene as nucleation seeds," *Nanoscale*, vol. 5, no. 12, pp. 5456–5461, 2013.
19. H. A. Chaliyawala, N. Rajaram, R. Patel, A. Ray, and I. Mukhopadhyay, "Controlled island formation of large-area graphene sheets by atmospheric chemical vapor deposition: role of natural camphor," *ACS Omega*, vol. 4, pp. 8758–8766, 2019.
20. W. Regan, et al., "A direct transfer of layer-area graphene," *Appl. Phys. Lett.*, vol. 96, no. 11, pp. 11–13, 2010.
21. F. Ravani, et al., "Graphene production by dissociation of camphor molecules on nickel substrate," *Thin Solid Films*, vol. 527, pp. 31–37, 2013.
22. M. Singh, H. S. Jha, and P. Agarwal, "Growth of large sp2 domain size single and multi-layer graphene films at low substrate temperature using hot filament chemical vapor deposition," *Mater. Lett.*, vol. 126, pp. 249–252, 2014.

Graphene-based photodetectors

4

Harsh A. Chaliyawala, Govind Gupta and Indrajit Mukhopadhyay

Contents

4.1 INTRODUCTION

A photodetector is an optoelectronic device that converts optical signals into electrical impulses. The fundamental operating principle of a metal-semiconductor photodetection device is based on the photoelectric effect. Unlike the pn-junction diode, the Schottky diode makes use of a metal-semiconductor junction for the rectifying function. The main advantage of using such devices is that they have very fast switching times due to their small capacitance and because they are carrier devices. Schottky diodes have a very short reverse recovery time, which is defined as the time needed for the diode to switch from the conducting to the non-conducting state as compared to pn-junction-based devices. For this reason, Schottky diodes are widely used in radio frequency (RF) circuits as mixers and detectors. A scheme illustrating the basic structure of metal/n-type

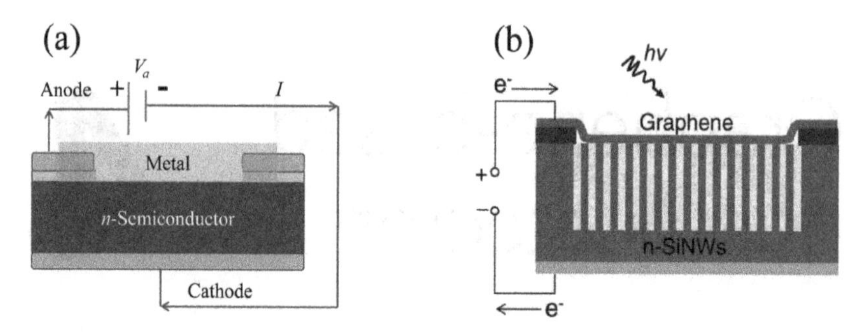

FIGURE 4.1 The basic structure of (a) a metal/n-semiconductor Schottky junction and (b) Gr/n-Si nanowire arrays (SiNWAs) photodetectors [20].

semiconductor and a metal/n-SiNWAs Schottky junction photodetector is shown in Figure 4.1(a). Under the unbiased condition, the built-in potential step and bent band features are similar to those of a semiconductor diode. When a positive or negative bias is applied, the bands shift in comparable ways. Therefore, a rectifying behavior can be expected in the metal-semiconductor junction. Initially, when a photon with energy greater than or equal to the bandgap energy interacts with the semiconducting material, the electrons present in the valence band come to an excited state by absorbing the energy from the incoming photons, thereby creating electron-hole (e-h) pairs. These photogenerated e-h pairs are then collected by their respective electrodes to accomplish current conduction. However, if the generated e–h pairs are not collected quickly, then they will recombine, giving up the extra energy in the form of photons. There is a possibility that the excited electron may jump from the valence band onto a trap state that exists within the bandgap. Further, if a photon has much higher energy, the bandgap energy enters the semiconducting material, then the electron may reach a high energy state in the conduction band, which then relaxes by thermalization process in which energy is released in the form of phonons and reaches the bottom of the conduction band.

Usually, planar Si substrates are not very beneficial due to the high specular reflection emanating from the surface, which is ~30%. As a result, keen interest in recent years has been focused on reducing the reflection loss emanating from the surface of Si by employing anti-reflection coatings such as Si_3N_4 and SiO_2. The thickness of the anti-reflection layer is optimized to a quarter of an incident wavelength, which makes the phase difference of the incident light and the reflected light equal to half a wavelength, thereby suppressing the reflection through destructive interference [1]. The fabrication process of the anti-reflection layer typically involves vacuum deposition equipment which makes the process costly. According to the new developments in this field, submicron sized holes on the surface allow light to travel inside the cavity and lead to in-plane light scattering, thereby enhancing strong photon absorption over a broad range of wavelengths [2]–[4]. In this regard, the fabrication of one-dimensional SiNWAs has drawn much attention in the field of energy conversion and optoelectronic devices in recent years [5], [6]. These SiNWAs are emerging as a promising option due to their varied electrical, optical, mechanical, and chemical properties which are the result of the confinement of photons, multiple reflections, and their unique dimensions [7]. In particular, the most interesting properties of SiNWAs include high-density electronic states [8], enhanced excitation binding energies and thermoelectric properties, high surface scattering for phonons and electrons, and high surface to volume ratio [9], [10], [11]. These outstanding properties can be coupled with graphene to form a Schottky junction as shown in Figure 4.1(b). Schottky junction–based photodetectors made from single layer and multilayer graphene sheets have been chosen over SiNWAs because of the high response at the desired wavelength. The development of such high-performance graphene/SiNWA Schottky junction photodiodes through novel and cost-effective fabrication processes has gained much interest. The working principle is similar to metal/*n*-semiconductor Schottky junctions. In this configuration, one-dimensional nanowire arrays work as an electron emitter, and photogenerated e-h pairs are collected according to their respective electrodes under the applied potential.

4.2 PARAMETERS UNDER EVALUATION

In this section, we discuss the performance parameters along with the physical properties that influence the performance of photodetectors. The various features of photodetectors such as spectral response, responsivity, dark current, quantum efficiency, forward-biased noise, noise equivalent power, capacitance, response time (rise and fall time), bandwidth, and cut-off frequency [11], [12] have applications in different fields.

1- Spectral response: A response that shows that the current is produced at a typical wavelength, and it is assumed that all wavelengths are at the same power density.

$$S_R = \frac{e \times \lambda}{h \times c} \quad (1)$$

where S_R is the spectral response of the device, e is electronic charge, λ is the incident wavelength, h is Planck constant, and c is the speed of light.

2- Quantum efficiency: It is the number of generated electron-hole pairs (i.e., current) divided by the number of photons.

$$\text{External quantum efficiency}\left(\text{EQE}\right)=\frac{J_{SC}}{\phi_{flux}\times e} \tag{2}$$

where J_{SC} = current of the generated charge carriers per unit area, ϕ_{flux} = photon flux, which can be described as

$$\phi_{flux}=\frac{\lambda\times P}{h\times c}=\frac{P}{E} \tag{3}$$

where P is the power incident at a particular wavelength and E is the photon energy.

3- Responsivity: Responsivity (R) is one of the important performance evaluation parameters that give information on the device responding to a particular wavelength of light (A/W). It is defined as the ratio of photocurrent (I_{Ph}) and the incident optical power intensity (P), where P_d is defined as the power density (P_d) and the active area of the device (A). The responsivity of a fabricated device is given by,

$$R=\frac{I_{Ph}}{P_d\times A} \tag{4}$$

4- Noise equivalent power (NEP): It is defined as the minimum amount of power of a given detector (of a given wavelength). The NEP expresses the sensitivity of the device and is given in Watts per square root of Hertz as:

$$\text{NEP}=\frac{\sqrt{2\times e\times I_D}}{R} \tag{5}$$

where R is the responsivity of the device.

5- Detectivity: It is defined as the inverse of NEP. The larger the detectivity of a photodetector, the more it is suitable for detecting weak signals which compete with the detector noise.

$$\text{Detectivity}\left(D\right)=R\times\sqrt{\frac{A}{2\times e\times I_D}}=\frac{\sqrt{A}}{\text{NEP}} \tag{6}$$

6- Terminal capacitance: It is the capacitance created between the p-n junction (metal-semiconductor Schottky junction) of the diode and the connectors of the device; it limits the response of the photodetector.

7- Response time: It is the time for the output signal to climb from 10% to 90% of its amplitude (rise time) and to drop from 90% to 10% (fall time).
8- Dark current: When a diode is in reverse bias and the electron-hole pair is generated in the absence of light, it is called dark current or leakage current.
9- Forward bias noise: It is a (current) source of noise that is related to the shunt resistance of the device. The shunt resistance is defined as the ratio of applied voltage (near 0V) to the amount of current generated. This is also called shunt resistance noise.
10- Frequency bandwidth: It is the frequency (or wavelength) range in which the photodetector is sensitive.
11- Cut-off frequency: It is the highest frequency at which the photodetector shows its sensitivity or it can be the smallest wavelength.

4.3 TRANSFER OF CAMPHOR-MADE GRAPHENE ON SiNWAs

In this section, we focus on the transfer of camphor etched MLG sheets onto SiNWAs to fabricate a Schottky junction–based photodetector. The scheme in Figure 4.2 illustrates the complete fabrication process of a device from APCVD camphor–derived MLG/SiNWAs via the Ag metal-assisted electroless chemical etching (MACE) technique. In step 1, the electroless etching consists of two mechanisms, which are based on the galvanic displacement reaction that occurs simultaneously at the Si/metal interface. In the first process, the Si substrate is immersed in an aqueous $HF/AgNO_3$ (catalytic agent) solution, and through an electrochemical redox reaction, as described in Eq 2(a–c), the Ag^+ ions adhere to the Si surface and act as the cathode, while the Si surface beneath acts as the anode [13]–[15]. Then Ag^+ ions capture the electrons from Si and form Ag nanoclusters. The electrons continue to transfer from Si to Ag nanoclusters and make Ag nanoclusters more negative, due to their high electronegativity, compared to Si.

Possible cathodic reaction:

$$2H^+ + 2e^- \rightarrow H_2 \qquad E_0 = +0.0\ V/NHE \tag{7a}$$

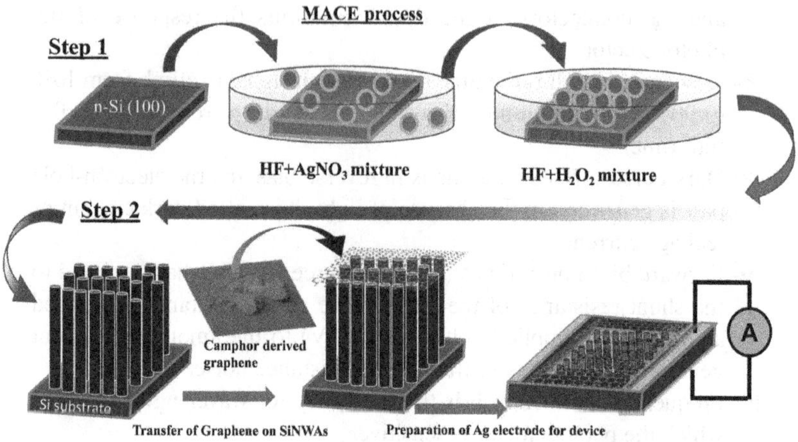

FIGURE 4.2 The synthesis process of vertical Si nanowire arrays (SiNWAs) using a metal-assisted chemical etching (MACE) process (step 1). Step 2 describes the transfer of a graphene sheet on the vertical SiNWAs through a polymer transfer process. The fabrication of the Gr/SiNWAs Schottky structure has been shown, followed by the preparation of front and back electrodes.

$$Ag^{+} + e^{-} \rightarrow Ag \qquad E_0 = +0.79\ V/NHE \tag{7b}$$

$$H_2O_2 + 2H^{+} \rightarrow 2H_2O + 2h^{+} \qquad E_0 = +1.76\ V/NHE \tag{7c}$$

Possible anodic reaction:

$$Si + 4h^{+} + 4HF \rightarrow SiF_4 + 4H^{+} \tag{8a}$$

$$SiF_4 + 2HF \rightarrow H_2SiF_6 \tag{8b}$$

$$Si + 4HF_2^{-} \rightarrow SiF_6^{2-} + 2HF + H_2 - +2e^{-} \qquad E_0 = -1.2\ V/NHE \tag{8c}$$

Overall reaction:

$$Si + H_2O_2 + 6HF \rightarrow 2H_2O + H_2SiF_6 + H_2^{-} - \tag{8d}$$

As the reaction progresses, Ag nanoparticles form nanoclusters, and excess oxidation occurs at the Si/metal interface under the influence of Ag nanoparticles. The Ag nanoparticles formed during the etching process are selectively removed by HF (Eq. 8(a-c)). When the processed Si substrate is dipped into an aqueous solution of HF/H_2O_2, a hole injection into Si occurs at the cathode site, since H_2O_2 acts as an oxidizing agent. At the anodic site, Si is oxidized and is dissolved in the Si/metal interface by HF (the possible reactions are described in Eq 8(d)), leading to the formation of vertically aligned one-dimensional silicon nanowires [16]–[18].

In step 2, after the formation of SiNWAs, the MLG sheet grown by camphor with the optimized conditions was transferred on a desired samples. The transfer was carried out by a lift off process using PMAA and, later, PMAA was removed in acetone [19]. The field emission scanning electron microscopy (FESEM) image of the surface morphology of the graphene sheet transferred onto SiNWAs shows a clear difference in the two regions, as observed in Figure 4.3(a). Moreover, the cross-sectional FESEM images in Figure 4.3(b) show an MLG sheet fully covered with highly dense and vertical SiNWAs suitable for the development of NIRPDs. This enables higher transparency leading to increased light absorption and carrier transport in the devices [20]–[22]. The homogeneous island growth formation of the MLG sheet is very well connected over the SiNWAs so that it functions as a transparent conductive sheet on the top surface of the semiconductor.

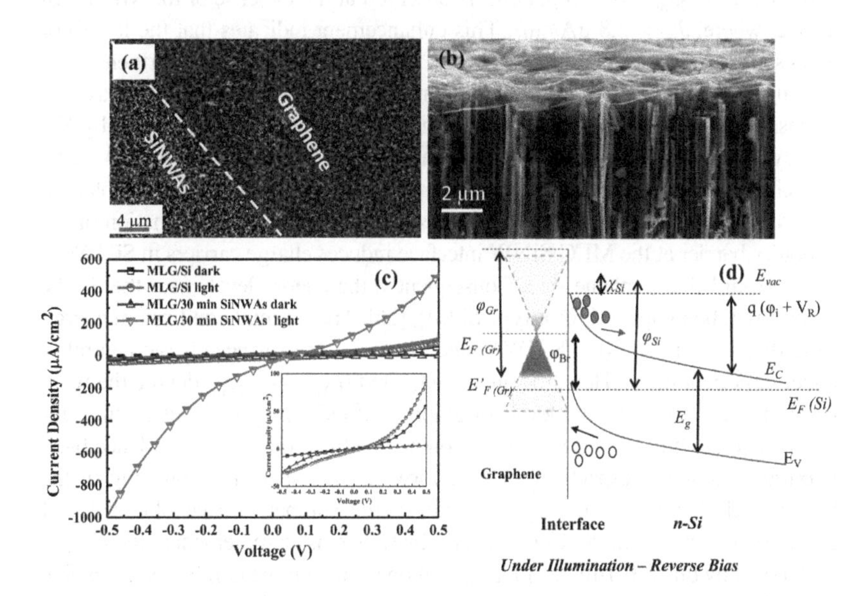

FIGURE 4.3 (a) Surface morphology of MLG grown on SiNWAs; (b) cross-sectional images of graphene on the top of SiNWAs; (c) J vs V curves under dark and light conditions for MLG/Si and MLG/ 30 min SiNWAs. Inset shows the zoomed-in image for the current values up to -50 μA/cm². (d) Schematic diagram of the energy band and charge carrier transport mechanism of the MLG/SiNWAs Schottky junction under the reverse bias condition.

4.4 PHOTORESPONSE PROPERTIES

This section describes the optoelectronic properties of MLG/SiNWAs photodetectors in comparison with MLG/Si Schottky junction–based photodetector. Figure 4.3(c) represents the typical current density-voltage (J-V) profile of MLG/Si and MLG/SiNWAs under dark and light conditions. Low rectifying behavior has been observed due to the formation of the Schottky contacts between MLG and Si. The dark current extracted is in few μA which is comparatively low at the expense of nanowire based devices (Inset of Figure 4.3(c)). This may be due to the lateral vibrations near the metal-semiconductor interface, thus developing more defect states and lowering the Schottky barrier height [23]. Thereby, electrons find an easier way to overcome the contact barrier under dark conditions, which ultimately affects the overall J-V performance. Under light conditions, it is observed that these devices exhibit pronounced rectification and photodiode characteristic effect upon incident radiation of 1000 W/cm^2. An enhancement in short circuit current density (J_{SC}) to 45.3 $\mu A/cm^2$ is observed at the expense of the MLG/n-Si device, where, $J_{SC} = 2.8$ $\mu A/cm^2$. This enhancement indicates that the length of NWs shows extreme support in the current transport mechanism. Nevertheless, the prolonged etching time results in the degradation of device performance as also discussed in our previous studies [24]. Similar behavior has been observed by Xie et al., with degradation of device performance above 20-min etched SiNWAs [25]. The charge transport mechanism at the interface of MLG/SiNWAs is explained schematically using energy band diagrams in Figure 4.3(d). The formation of the Schottky barrier at the MLG/SiNW interface induces charge carriers in SiNWs to move toward the graphene side. Consequently, the energy levels near the SiNW surface will bend upward for n-type Si [21], [26]. The thermal equilibrium energy band diagram of the MLG/SiNWA junction shows the surface charge neutrality under dark conditions. Here, the (φ_{MLG})/(φ_{Si}) and ($E_{F(MLG)}$)/($E_{F(Si)}$) denote the work functions and Fermi energy levels of graphene/(SiNWs). χ_{Si} is the electron affinity of silicon, while E_C and E_V are the conduction band minima and valence band maxima of silicon, respectively. Under forward bias, there is a lowering of the Fermi level of graphene ($E_{F\,(MLG)}$); it reduces the number of accessible states of photoexcited holes from Si. Under reverse bias, there is an increment in the $E_{F(MLG)}$ as denoted by quasi-Fermi level $E'_{F\,(MLG)}$; it opens up a large number of accessible states that can be occupied by photoexcited holes injected from Si. Thus a large photocurrent under the reverse bias condition has been observed.

In order to understand the mechanism of MLG/Si and MLG/SiNWA devices upon light illumination, time-correlated photoresponse characteristics and various properties have been investigated from the fabricated devices as shown in Figure 4.4. The devices are found to be highly responsive toward photons having

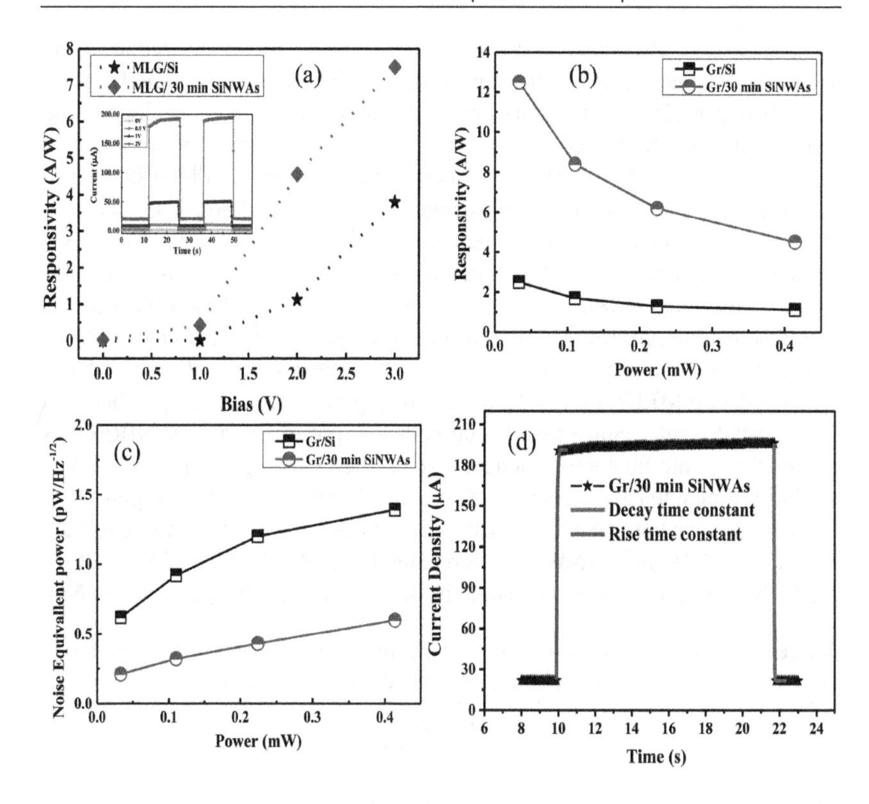

FIGURE 4.4 (a) Bias-dependent photoresponsivity study of MLG/Si and MLG/30 min SiNWAs. The inset shows the time-correlated photoresponse for MLG/30 min SiNWAs Schottky junction NIRPDs. The graphs show (b) the influence of optical power intensity on responsivity (c) NEP for the MLG/Si and MLG/30 min SiNWAs based NIRPDs. (d) Rise time and decay time fitted curves for response time constants of the MLG/ 30 min SiNWAs based NIRPDs.

a wavelength of 940 nm which has been demonstrated by the time-dependent photoresponse measurements carried out at different bias variations from V_{bias} = 0, 1, 2, and 3 V as plotted in Figure 4.4(a). When the light was turned ON and OFF alternatively, the MLG/Si and MLG/ 30 min SiNWA devices show a photoresponse under different bias values (Inset of Figure 4.4(a)). The ON-OFF switching cycle shows a good stability and repeatability with high photoresponse. A linear dependence of the applied external bias with the photocurrents is observed with a four-fold enhancement in the MLG/SiNWA-based NIRPDs as compared to the MLG/Si-based device. At 0V, the I_{ph} values of 36 μA/W and 22.1 mA/W, for MLG/Si and MLG/SiNWAs, respectively, have been obtained. The photocurrent enhancement upon illumination under the no bias condition can be attributed to have a sufficient built-in potential (V_{bi}) at the interface of the graphene/Si junction which enhances the charge carrier separation in the device. The enhancement in the current for the MLG/SiNWA photodetectors at 0V bias can also be due to its broadband light absorption capability at the expense of MLG/Si photodetectors, which accelerate more electron-hole pair generation leading to high carrier collection with the help of a built-in electric field [22], [32]. A device that can operate at 0V bias condition is known as a self-driven photodetector. As we increase the bias to 1, 2, and 3 V, there is a tremendous enhancement in the photocurrent for MLG/SiNWAs as compared to MLG/Si-based NIRPDs. There are various graphene/Si-based photodetectors which show promising results in terms of responsivity and detectivity as compared to our present work (Table 4.1). It is clear that the *R* value of our camphor-based MLG/SiNWA device are very promising and comparable with the reported results. Since fast detection at low power signals for NIRPDs is desired for various photodetector applications, the switching is performed at low to high power intensities from ~33.7 μW to 0.4 mW as shown in Figure 4(b). Interestingly, for lower power intensity the *R* value shows a high value of 12.5 A/W for MLG/30 min SiNWAs as compared to 2.5 A/W for MLG/Si NIRPDs due to the high surface to volume ratio leading to excellent light trapping in NW

TABLE 4.1 Comparison of the device performance of the NIRPDs fabricated in the present study and other photodetectors having similar structures.

DEVICE STRUCTURE	*RESPONSIVITY (A/W)*	*RESPONSE/ RECOVERY TIME CONSTANTS (MS)*	I_L/I_D	*DETECTIVITY (J)*	*REF*
MLG/n-Si (camphor based)	1.12	69.1/69.5	~ 10^1	~10^{12}	[24]
MLG/30 min n-SiNWAs (camphor based)	4.56	34/13	~10^1	~10^{13}	[24]
MLG/n-Si	0.43	1.2/3	~10^4	~10^8	[27]
MLG/p-Si	$9.9E^{-3}$	/	/	/	[28]
MLG/pure SiNWs	/	20	~10^3	/	[29]
MLG/Si waveguide	0.13	/	/	/	[30]
MLG/CH_3–Si nanohole array	0.38	/	/	~10^{13}	[31]

structures. Furthermore, Figures 4(c) and (d) show the effect on NEP at various optical power intensities and the rise time and decay time constants associated with the representative devices. MLG/30 min SiNWAs is found to have a very low NEP of 0.6 pW $Hz^{-1/2}$ compared to a relatively high NEP of 1.4 pW $Hz^{-1/2}$ for MLG/Si-based NIRPDs at 2V bias [32], [33]. The low NEP signifies that the device yields a faster and highly sensitive photoresponse even from high-power intensity signals. Moreover, the rise and decay time constants evaluated from the fitted data were found to be 34 and 13 ms, respectively, which shows very fast switching behavior in the NIRPDs designed in the present study.

4.5 CONCLUSION

In summary, high-performance NIRPDs have been successfully fabricated by using a camphor-based MLG sheet transferred on n-Si and SiNWAs. The single crystalline structure and the number of layers for MLG was confirmed by TEM and SAED patterns and also validated by Raman mapping analysis. The device with planar MLG/n-Si and MLG/n-SiNWAs showed pronounced photovoltaic behavior under illumination conditions. The optimization of the maximum absorption capability of NW arrays and increase in the surface recombination centers for longer NW length leads to a maximum J_{SC} of ~40 μA cm^{-2} for MLG/30 min SiNWA photodetectors. The responsivity and detectivity at zero bias voltage were estimated to be ~22.1 mAW^{-1} and ~9.2×10^{11} Jones for MLG/30 min SiNWA photodetectors compared to ~36 μAW^{-1} and ~4.5×10^{9} Jones for planar MLG/Si photodetectors, respectively. The highest responsivity of 12.5 AW^{-1} has been achieved for MLG/30 min SiNWAs at a very low power signal of 33 μW. Moreover, device analysis shows that MLG/30 min SiNWA photodetectors can work at a high response speed of 34/13 ms and with very low NEP of 0.60 pW $(Hz)^{-1/2}$ at the expense of planar MLG/Si devices. Therefore, the facile and low-cost development of photodetectors makes it a promising candidate for future high-performance SiNWA-based NIRPDs.

REFERENCES

1. H. K. Raut, V. A. Ganesh, A. S. Nair, and S. Ramakrishna, "Anti-reflective coatings: a critical, in-depth review," *Energy Environ. Sci.*, vol. 4, no. 10, pp. 3779–3804, 2011.

2. Z. Yu, A. Raman, and S. Fan, "Fundamental limit of nanophotonic light-trapping in solar cells," *Next Gener. Photonic Cell Technol. Sol. Energy Convers.*, vol. 7772, p. 77720Z, 2010.
3. J. Tang, J. Shi, L. Zhou, and Z. Ma, "Fabrication and optical properties of silicon nanowires arrays by electroless Ag-catalyzed etching," *Nano-Micro Lett.*, vol. 3, no. 2, pp. 129–134, 2011.
4. Y. F. Huang, S. Chattopadhyay, Y. J. Jen, C. Y. Peng, T. A. Liu, Y. K. Hsu, Ci-L. Pan, H. C. Lo, C. H. Hsu, Y. H Chang, C. S Lee, K. H. Chen, and L. C. Chen, "Improved broadband and quasi-omnidirectional anti-reflection properties with biomimetic silicon nanostructures," *Nat. Nanotechnol.*, vol. 2, no. 12, pp. 770–774, 2007.
5. L. Cao, P. Fan, A. P. Vasudev, J. S. White, Z. Yu, W. Cai, J. A. Schuller, S. Fan, and M. L. Brongersma, "Semiconductor nanowire optical antenna solar absorbers," *Nano Lett.*, vol. 10, no. 2, pp. 439–445, 2010.
6. S. A. Razek, M. A. Swillam, and N. K. Allam, "Vertically aligned crystalline silicon nanowires with controlled diameters for energy conversion applications: experimental and theoretical insights," *J. Appl. Phys.*, vol. 115, no. 19, p. 194305, 2014.
7. V. Schmidt, J. V. Wittemann, S. Senz, and U. Gósele, "Silicon nanowires: a review on aspects of their growth and their electrical properties," *Adv. Mater.*, vol. 21, no. 25–26, pp. 2681–2702, 2009.
8. H. Scheel, S. Reich, and C. Thomsen, "Electronic band structure of high-index silicon nanowires," *Phys. Status Solidi Basic Res.*, vol. 242, no. 12, pp. 2474–2479, 2005.
9. M. Paniccia, "Integrating silicon photonics," *Nat. Photonics*, vol. 4, no. 8, pp. 498–499, 2010.
10. S. F. Bahram Jalali, "Silicon photonics," *J. Light WAVE Technol.*, vol. 24, no. 12, pp. 381–429, 2008.
11. X. Li, et al., "High detectivity graphene-silicon heterojunction photodetector," *Small*, vol. 12, no. 5, pp. 595–601, 2016.
12. C. Xie, Y. Wang, Z. X. Zhang, D. Wang, and L. B. Luo, "Graphene/semiconductor hybrid heterostructures for optoelectronic device applications," *Nano Today*, vol. 19, pp. 41–83, 2018.
13. W. Q. Xie, J. I. Oh, and W. Z. Shen, "Realization of effective light trapping and omnidirectional antireflection in smooth surface silicon nanowire arrays," *Nanotechnology*, vol. 22, no. 6, 2011.
14. S. Kato, et al., "Optical assessment of silicon nanowire arrays fabricated by metal-assisted chemical etching," *Nanoscale Res. Lett.*, vol. 8, no. 1, pp. 1–6, 2013.
15. S. K. Srivastava, D. Kumar, P. K. Singh, M. Kar, V. Kumar, and M. Husain, "Excellent antireflection properties of vertical silicon nanowire arrays," *Sol. Energy Mater. Sol. Cells*, vol. 94, no. 9, pp. 1506–1511, 2010.
16. Z. Huang, N. Geyer, P. Werner, J. De Boor, and U. Gösele, "Metal-assisted chemical etching of silicon: a review," *Adv. Mater.*, vol. 23, no. 2, pp. 285–308, 2011.
17. M. L. Zhang, et al., "Preparation of large-area uniform silicon nanowires arrays through metal-assisted chemical etching," *J. Phys. Chem. C*, vol. 112, no. 12, pp. 4444–4450, 2008.

18. X. Zhong, Y. Qu, Y. C. Lin, L. Liao, and X. Duan, "Unveiling the formation pathway of single crystalline porous silicon nanowires," *ACS Appl. Mater. Interfaces*, vol. 3, no. 2, pp. 261–270, 2011.
19. W. Regan, et al., "A direct transfer of layer-area graphene," *Appl. Phys. Lett.*, vol. 96, no. 11, pp. 11–13, 2010.
20. G. Fan, et al., "Graphene/silicon nanowire Schottky junction for enhanced light harvesting," *ACS Appl. Mater. Interfaces*, vol. 3, no. 3, pp. 721–725, 2011.
21. X. Zhang, C. Xie, J. Jie, X. Zhang, Y. Wu, and W. Zhang, "High-efficiency graphene/Si nanoarray Schottky junction solar cells via surface modification and graphene doping," *J. Mater. Chem. A*, vol. 1, no. 22, pp. 6593–6601, 2013.
22. Y. Wu, et al., "Graphene transparent conductive electrodes for highly efficient silicon nanostructures-based hybrid heterojunction solar cells," *J. Phys. Chem. C*, vol. 117, no. 23, pp. 11968–11976, 2013.
23. N. Wang, Y. H. Tang, Y. F. Zhang, C. S. Lee, I. Bello, and S. T. Lee, "Si nanowires grown from silicon oxide," *Chem. Phys. Lett.*, vol. 299, no. 2, pp. 237–242, Jan. 1999.
24. H. Chaliyawala, N. Aggarwal, Z. Purohit, R. Patel, G. Gupta, A. Jaffre, S. Le Gall, A. Ray, and I. Mukhopadhyay, "Role of nanowire length on the performance of self-driven NIR photodetector based on mono/bi-layer graphene (camphor)/Si-nanowire Schottky junction," *IOP Nanotechnol.*, vol. 31, p. 225208, 2020.
25. C. Xie, et al., "Monolayer graphene film/silicon nanowire array Schottky junction solar cells," *Appl. Phys. Lett.*, vol. 99, no. 13, pp. 2011–2014, 2011.
26. X. Miao, et al., "High efficiency graphene solar cells by chemical doping," *Nano Lett.*, vol. 12, no. 6, pp. 2745–2750, 2012.
27. X. An, F. Liu, Y. J. Jung, and S. Kar, "Tunable graphene—silicon heterojunctions for ultrasensitive photodetection," vol. 13, no. 3, pp. 909–916, 2013.
28. M. Amirmazlaghani, F. Raissi, O. Habibpour, J. Vukusic, and J. Stake, "Graphene-Si Schottky IR detector," *IEEE J. Quantum Electron.*, vol. 49, no. 7, pp. 589–594, 2013.
29. C. Xie, et al., "Schottky solar cells based on graphene nanoribbon/multiple silicon nanowires junctions," *Appl. Phys. Lett.*, vol. 100, no. 19, p. 193103, 2012.
30. X. Wang, Z. Cheng, K. Xu, H. K. Tsang, and J. Bin Xu, "High-responsivity graphene/silicon-heterostructure waveguide photodetectors," *Nat. Photonics*, vol. 7, no. 11, pp. 888–891, 2013.
31. L. Zeng, et al., "Bilayer graphene based surface passivation enhanced nano structured self-powered near-infrared photodetector," *Opt. Express*, vol. 23, no. 4, p. 4839, 2015.
32. L. B. Luo, et al., "Light trapping and surface plasmon enhanced high-performance NIR photodetector," *Sci. Rep.*, vol. 4, 2014.
33. N. Aggarwal, et al., "A highly responsive self-driven UV photodetector using GaN nanoflowers," *Adv. Electron. Mater.*, vol. 3, no. 5, pp. 1–7, 2017.

A Novel Graphene as an Anodic Material for Lithium-Ion Battery

5

Kashinath Lellala, Harsh A. Chaliyawala and Indrajit Mukhophadhyay

Contents

5.1 INTRODUCTION

Energy-related crises have already hampered the development and modernization of society. Hence, the engineering and research community are engaged in the exploration of sustainable energy resources and conversion systems such as fuel cells, supercapacitors, solar cells, and batteries [1–4]. Among the energy-related resources, lithium-ion batteries (LIBs) are representative of rechargeable batteries (capacity of 700–4500 mAh with an operating voltage of 3.6 V) having superior properties such as high energy density, low self-discharging, and confined memory. LIBs are lightweight and have been widely used in portable electronic devices such as notebooks, cell phones, digital cameras, and computers [5–8]. The operation mechanism of LIBs is based on the commuting movements of lithium ions through the electrodes with high physicochemical properties. A promising electrode material is required for the high performance of LIBs and for controlling the process of lithiation/delithiation, which is the objective of the present research. In general, an anode material should possess three distinct characteristics and physicochemical behavior such as (i) intercalation-deintercalation, (ii) alloying-dealloying, and (iii) redox conversion for the performance of LIBs. Fabrication of high-performance and higher-efficiency anodic materials is needed for the development of LIBs with high energy density for all electronic and electrical devices. The advantages of LIBs are that they possess high energy density, have long cyclic life, exhibit high coulombic energy efficiency and flexibility in designing, and are helpful in developing a wide range of electronic products. In general, the charge/discharge tendency of LIBs at 0.2–1°C takes approximately 5–1 h for the full capacity of the cell for storage or utilization at the usual operating temperature ranging from 15 to 60°C. If the temperature is less than 15°C, the capacity becomes low; when the temperature is greater than 60°C, slow degradation of the electrolyte and electrode materials occurs after a certain period of time. These are the major problems to be considered for improving LIBs performance. During charging/discharging, Li-ions are shuttled between the anode/cathode through the electrolyte, the electrons are changed through the external circuit electrons through the external circuit; the electrons and Li-ions reach the active sites of the electrode materials and complete the energy transformation between chemical and electrical energy. In particular the Li^+ diffusion time within the electrode mainly depends upon the diffusion path length (L) and diffusion coefficient (D) which relates with $T = L^2/D$ where L depends upon the size and dimensions of the electrode material, and D represents the movement or drift of Li^+. Reduction in L or increase in D can improve Li^+ diffusion, which depends upon the quality and quantity of the anode materials, and nano-level dimensions can also improve the electrochemical performance [9–10].

Due to the poor electrical conductivity of the electrode materials and lithium diffusion, the performance and efficiency of LIBs electrode materials are degraded or lowered. This requires lower-sized nanostructured particles, and their combination with high or active conductive materials, such as carbon-based materials, has played an important role in current research trends [11]. Dendrite growth and Li plating are the major problems caused during charging/discharging of high-scale power batteries. The mechanism of the disordered structure of nano-carbon materials for Li-ion storage is complex, but it clearly depends on the characteristic of amorphous materials. Thus researchers are interested in the development of micro or nanocrystalline graphene layers bonded with van der Waals forces[11–13]. Massless Dirac fermions and sheet-like (2D) graphene derivatives have great potential due to their outstanding physicochemical properties for battery applications. The discovery of graphene with its fascinating properties has given rise to a new class of two-dimensional materials. Graphene has the advantage of a layered structure, and its unique physical and chemical properties have become the focus of intense research. Graphene-based materials have remarkable physicochemical properties that are better than those of pure nanomaterials, and they also possess excellent cyclic performance, high stable reversible capacity, and high rate capability, providing a surplus porous channel for faster diffusion of Li-ions and in turn preventing the agglomeration of nanoparticles. Graphene-related materials which include graphene sheets, single-layered graphene (SLGR) sheets and few-layered graphene (FLGR) sheets, reduced graphene oxide, and graphene oxide possess multifunctional physiochemical properties. These properties help in building nanostructured materials with enhanced physicochemical properties, which have contributed to many research applications [14]. The electrolyte/solvent intercalation and the dispersive reaction of graphene layers cause the capacitive behavior of disordered carbon with high rate performance and low coulombic efficiency. Koratkar's research group showed that the performance of graphene (Gr) in the polymer matrix is better than carbon nanotubes (CNT) in terms of mechanical properties [15]. A maximum capacity of 740 $mAhg^{-1}$ adsorption for double-layered graphene sheets was calculated theoretically [16]. Sato et al. reported on the disordered carbons limit in Li covalent bonding sites which are trapped in benzene rings showing a reversible capacity of 1116 $mAhg^{-1}$ [17]. Uthaisar studied the edge effects of Li diffusion in graphene by density functional theory (DFT) and additionally showed that graphene nano ribbons with zigzag structure and morphology facilitate a fast discharge rating due to the decrease in diffusion length and the lowering of energy barriers [18]. Xiao et al. reported the hierarchical structures of functionalized graphene sheets with the highest discharge capacity of 15,000 $mAhg^{-1}$ at a current density of 0.1 mA cm^2 [19]. Guo et al. reported crumpled paper-like graphene nano-sheets obtained from oxidation, rapid thermal expansion, and ultra-sonication treatment of graphite, which showed a superior capacity of 672 $mAhg^{-1}$ with good cyclic stability [20]. Wang et al. reported a discharge capacity of 945 $mAhg^{-1}$ and a charge capacity 700 $mAhg^{-1}$ for a flower-shaped reduced graphene oxide (RGO) structure [21]. Moreover, Graphene developed on a Cu foil by the CVD method showed a maximum energy density of 10 Whl^{-1} [22]. 3D graphene developed using the hydrothermal method exhibited a capacity of 170 $mAhg^{-1}$ at 145 mA g^{-1} and stretchable reduced graphene oxide fabricated from vacuum filtration exhibited a capacity of 1393 mA h g^{-1} [23]. Zhao et al. synthesized flexible graphene by mechanical cavitation and the oxidation method showed a reversible capacity of 300 mA h g^{-1} and excellent electrode stability even after 400 cycles [24]. Lian et al. studied FLGR sheets with a first cycle capacity of 1264 mA h g^{-1} at 100 mA h g^{-1} and the capacity decayed to 848 mA h g^{-1} after 40 cycles at a current density of 100 mA g^{-1} [25]. Zhang et al. studied graphene foam with 340 mA h g^{-1} at 100 mA g^{-1} [26]. Liu et al. synthesized interconnected, folded porous graphene aerogels using a mechanical pressing process exhibiting a high capacity of 1091 and 864 mA h g^{-1} and the reversible capacity after 100 cycles was maintained at 568 mA h g^{-1} [27]. Guo et al. reported on the fabrication of hollow graphene oxide spheres using a water and oil emulsion method with a reversible capacity of 485 mA h g^{-1} [28]. In this chapter, we have successfully synthesized graphene films on a copper substrate from camphor for LIB applications. Single- and bi-layered graphene sheets were examined for LIB applications. The charge/discharge capacity and lithium intercalation with graphene layers were examined, and the electrochemical studies are summarized in detail.

5.2 EXPERIMENTAL METHODOLOGY

The synthesis and characterization of graphene layers on a copper substrate (Cu (111)) from camphor are summarized in Chapter 2. The detailed characterizations of the identification of graphene have been evaluated by Raman and HRTEM images and other characterizations of synthesized graphene from camphor.

5.3 ELECTROCHEMICAL STUDIES

Electrochemical studies were performed using a three-electrode configuration in a standard swage lock cell assembled inside an Ar glove box by maintaining the oxygen and water content below 1 ppm. A Cu-coated graphene electrode was used as the working electrode, Li foil as the counter, and a Li-tipped Cu rod as the reference electrode. The working and counter electrodes were separated by polyethylene Celgard 2400. The electrolyte was 1.0 M $LiPF_6$ dissolved in a mixture of ethylene carbonate (EC), ethylmethyl carbonate (EMC), and dimethyl carbonate (DMC) (1:1:1, v/v). Battery performance and electrochemical impedance spectroscopy studies were conducted using the biologic BT-20 electrochemical workstation at room temperature with 10 mV amplitude in the range of 100 kHz to 1 Hz in automatic sweep mode from high to low frequencies.

5.4 RESULTS AND DISCUSSION

5.4.1 Cyclic voltammetry

Cyclic voltammetry was performed to understand the underlying mechanism, chemical interaction, and electrochemical properties of SLGR and FLGR sheets in a potential window of 0–3 V vs Li^+/Li at a scan rate of 0.5 mV/s as shown in Figure 5.1(a). The cyclic voltammetry profile shows a sharp and well-resolved peak that signifies the Li^+ insertion/disinsertion at a faster rate, whereas a broad peak suggests a sluggish process due to higher Li diffusion rate in SLGR than the FLGR. The cathodic sweep shows two irreversible peaks at 0.67 V and 0.79 V for SLGR. FLGR shows two irreversible cathodic peaks at 0.45 V and 0.6 V. These peaks are attributed to the stepwise formation of solid electrolyte interface (SEI) layers from the side of the reaction of electrolytes on the graphene electrolyte interface and the intercalation/insertion of Li^+ on/into the graphene layer structure, respectively. However, the intensity level is higher for SLGR than FLGR which predicts the nature of higher activity and defect-free bonding structure of SLGR. The anodic scan shows two broad peaks at 1.05 V and 1.25 V for SLGR, which can be attributed to the deintercalation and oxidation of electroactive species on SEI. However, for FLGR a peak at 1.2 V shows perfect delithiation. From cyclic voltammetry, it is clearly evident that the redox nature of SLGR is very stronger than FLGR and double-layer capacitor behavior is higher for FLGR, which can be attributed to the defect nature and island formation of FLGR. Additionally, the nature of interaction with lithium diffusion or drifting is very less in SLGR than in FLGR. The faster Li^+ insertion is attributed to the transformation and stable defect-free surface. The potential of graphene indicates smaller polarization and better rate capabilities. According to cyclic voltammetry, the area

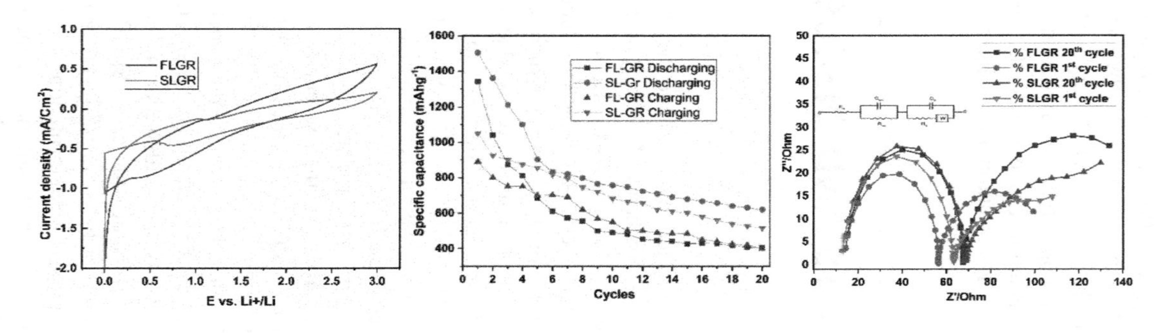

FIGURE 5.1 (a) Cyclic voltammetry, (b) discharging/charging capacities till the 20th cycle, and (c) EIS spectra of few-layer graphene (FLGR) sheets and single-layer graphene (SLGR) sheets.

of oxidation peaks is smaller than that of reduction peaks, thereby implying the excellent reversible nature of graphene films. The double-layer capacitor behavior is stronger for FLGR than SLGR due to higher oxidation and low reduction potential when compared to SLGR [29–31].

5.4.2 Charge/discharging performance

The electrochemical properties of the graphene film (SLGR and FLGR) electrodes were evaluated by galvanostatic discharge/charge measurements at room temperature. The first discharge capacity (lithium insertion) was 1540 $mAhg^{-1}$ and the first reversible charge (lithium extraction) capacity was 1050 $mAhg^{-1}$ at a current density of 0.1 mA cm^{-2} for the SLGR film with a specific reversible capacity retention of 50% till 20 cycles. The first reversible discharging and charging for FLGR films were 1340 $mAhg^{-1}$ and 890 mA h g^{-1} at a current density of 0.1 mA cm^{-2} with a retention of 45% reversible charge capacity till 20 cycles (Figure 5.1(b)). From the charge/discharge cycling performance, we determine the ability of each cycle in which the first cycle of SLGR films has a larger plateau than FLGR films which depends on the island formation of graphene and defect-free structure of carbon bonding [32–34]. The initial capacity is the highest and the process of discharge is the longest. For the first five cycles, the discharge capacity gradually decreased, but after the sixth cycle, it showed a stable and very slow decrease for SLGR. Similarly, the charging ability of SLGR shows a standard charging attitude up to the 20th cycle, and the performance is one-fourth of the initial charging capacity. The discharging of the first five cycles shows a decrement of capacity up to one-third of the initial value, and after the sixth cycle, the decrease of discharge capacity is gradual in nature and attains one-fourth of the initial discharging/charging capacity even after the 20th cycle for FLGR. There are not many drastic decreases in charging capacity for both SLGR and FLGR. The discharge plateau for the second and third is almost similar, which indicates that the battery structure of the solid electrolyte interface is stable and the layered structure has protected the graphene film. A linear voltage relationship has been observed due to the formation of electrical double-layer capacitors or pseudocapacitors [35,36]. The graphene film deposited on copper substrates from camphor thus showed very good performance for LIBs.

5.4.3 Electrochemical impedance spectroscopy

Electrochemical measurements were performed in a two-electrode system, and the counter electrode is a lithium foil about 1.0 cm^2. The counter electrode is large enough, and so it almost does not affect the EIS behavior. The radius of the semicircle increases with the increase in the cycle, and the difference in the radius of the first cycle is less than the radius of the 20th cycle for SLGR and FLGR revealing the formation of passivating films during the first charge/discharge process, and the latter shows an almost stable performance of discharge/charge as shown in Figure 5.1(c). It can be inferred from the increasing cyclic performance that the state of charge has more influence on the impedance at the first discharge/charge cycle [36,37]. Nyquist plots provide information on the resistances of the electrolyte [8], surface film, and charge-transfer process. All Nyquist plots showed significant overlap and similar charge-transfer resistances, indicating that the electrode reaction kinetics remained largely unchanged during cycling. An elongated semicircle region of the controlled kinetic of the charge-transfer region at high frequencies and a straight line corresponding to the mass-transfer region were controlled by Warburg impedance at lower frequencies [38–41]. An equivalent circuit analysis is illustrated in the inset of Figure 5.1(c), where R_s denotes the electrolyte resistance, R_{ct} denotes the resistance of electrode and electrolyte due to the charge-transfer, the constant phase element (CPE) is related to the double-layer capacitor, and Z_w is the result of diffusion of Li^+ known as the Warburg impedance (*W*), which is the combined effect of diffusion of electrolytes and electrode interfaces generating as a straight line at a lower frequency. The combination of charge-transfer impedance and Warburg impedance is known as Faradic impedance which indicates the kinetic of the cell reaction and the state of lithiation of graphene films [42–45]. The R_s is attributed to ion transport inside the electrolyte and separator; RSEI-CPE_1 confirms the lithium-ion shifting through SEI layers; RCT and CPE_2 indicate the charge-transfer at the interface of electrolytes/electrodes, and *W* is due to lithium-ion diffusion inside the bulk phase.

5.5 CONCLUSION

In conclusion, successful synthesis of single and bi-layer graphene films on a copper substrate from camphor was examined for LIB applications. The efficiency and robust performance of graphene layers and their intercalation with lithium are summarized in detail. The flexible nature of graphene sheets can enhance the lesser volume expansion effectively. The results signify better performance of SLGR than FLGR due to perfect island formation and better defect structure of carbon bonding of SLGR than FLGR.

ACKNOWLEDGMENT

The authors thank DST (SERB), Govt. of India, for financial support of the current work (Project Nos. DST/TMD/MES/2K17/32(G) and SERB/CRG/2018/002067).

REFERENCES

1. S. Han, D. Wu, S. Li, F. Zhang, and X. Feng, "Porous graphene materials for advanced electrochemical energy storage and conversion devices," *Adv. Mater.*, vol. 26, pp. 849–864, 2014.
2. L. Qie, W.-M. Chen, Z.-H. Wang, Q.-G. Shao, X. Li, L.-X. Yuan, X.-L. Hu, W.-X. Zhang, and Y.-H. Huang, "Nitrogen-doped porous carbon nanofiber webs as anodes for lithium ion batteries with a superhigh capacity and rate capability," *Adv. Mater.*, vol. 24, pp. 2047–2050, 2012.
3. G. Q. Zhang, and X. W. Lou, "General solution growth of mesoporous $NiCo_2O_4$ nanosheets on various conductive substrates as high-performance electrodes for super capacitors," *Adv. Mater.*, vol. 25, pp. 976–979, 2013.
4. Z. Y. Wang, Y. F. Dong, H. J. Li, Z. B. Zhao, H. B. Wu, C. Hao, S. H. Liu, J. S. Qiu, and X. W. Lou, "Enhancing lithium–sulphur battery performance by strongly binding the discharge products on amino-functionalized reduced graphene oxide," *Nat. Commun.*, vol. 5, p. 5002, 2014.
5. M. Wakihara, and O. Yamamoto, Eds. *Li-Ion Batteries: Fundamentals and Performance*; Wiley-VCH: New York, 1998.
6. G.-A. Nazri, and G. Pistoia, Eds. *Lithium Batteries: Science and Technology*; Kluwer Academic Publishers: New York, 2003.
7. K. Ozawa, Ed. *Lithium Ion Rechargeable Batteries*; Weinheim, Germany: Wiley-VCH, 2009.
8. K. E. Alifantis, S. A. Hackney, and R. Vasant Kumar, Eds. *High Energy Density Lithium Batteries: Materials, Engineering, Applications*; Weinheim, Germany: Wiley VCH, 2010.
9. Y. Tang, Y. Zhang, W. Li, B. Ma, and X. Chen, "Rational material design for ultrafast rechargeable lithium-ion batteries," *Chem. Soc. Rev.*, vol. 44, p. 5926, 2015.
10. H. Ren, J. Sun, R. Yu, M. Yang, L. Gu, P. Liu, H. Zhao, D. Kisailus, and D. Wang, "Controllable synthesis of mesostructures from TiO_2 hollow to porous nanospheres with superior rate performance for lithium ion batteries," *Chem. Sci.*, vol. 7, p. 793, 2016.
11. H. Chen, M. B. Muller, K. J. Gilmore, G. G. Wallace, and D. Li, "Mechanically strong, electrically conductive, and biocompatible graphene paper," *Adv. Mater.*, vol. 20, pp. 3557–3561, 2008.

12. D. Aurbach, *Advances in Lithium-Ion Batteries*, Eds. W. A. van Schalkwijk, and B. Scrosati; Springer: Boston, MA, 7–77, 2002.
13. L. Cheng, X.-L. Li, H.-J. Liu, H.-M. Xiong, P.-W. Zhang, and Y.-Y. Xia, "Carbon-Coated Li4Ti5O12 as a High Rate Electrode Material for Li-Ion Intercalation," *J. Electrochem. Soc.*, vol. 154, pp. A692–A697, 2007.
14. D. Aurbach, "Review of selected electrode–solution interactions which determine the performance of Li and Li ion batteries," *J. Power Sources*, vol. 89, pp. 206–218, 2000.
15. E. Buiel, A. E. George, and J. R. Dahn, Li-insertion in hard carbon anode materials for Li-ion batteries," *J. Electrochem. Soc.*, vol. 145, pp. 2252–2257, 1998.
16. J. R. Dahn, T. Zheng, Y. H. Liu, and J. S. Xue, "Mechanisms for Lithium Insertion in Carbonaceous Materials," *Science*, vol. 270, pp. 590–593, 1995.
17. K. Sato, M. Noguchi, A. Demachi, N. Oki, and M. Endo, "A Mechanism of Lithium Storage in Disordered Carbons," *Science*, vol. 264, pp. 556–558, 1994.
18. C. Uthaisar, and V. Barone, "Edge effects on the characteristics of li diffusion in grapheme," *Nano Lett.*, vol. 10, pp. 2838–2842, 2010.
19. J. Xiao, D. Mei, X. Li, W. Xu, D. Wang, G. L. Graff, W. D. Bennett, Z. Nie, L. V. Saraf, I. A. Aksay, J. Liu, and J.-G. Zhang, "Hierarchically porous graphene as a lithium–air battery electrode, " *Nano Lett.*, vol. 11, pp. 5071–5078, 2011.
20. P. Guo, H. Song, and X. Chen, "Electrochemical performance of graphene nanosheets as anode material for lithium-ion batteries," *Electrochem. Commun.*, vol. 11, pp. 1320–1324, 2009.
21. G. X. Wang, X. P. Shen, J. Yao, and J. Park, "Graphene nanosheets for enhanced lithium storage in lithium ion batteries," *Carbon*, vol. 47, pp. 2049–2053, 2009.
22. D. Wei, S. Haque, P. Andrew, J. Kivioja, T. Ryhanen, A. Pesquera, A. Centeno, B. Alonso, A. Chuvilin, and A. Zurutuza, "Ultrathin rechargeable all-solid-state batteries based on monolayer grapheme," *J. Mater. Chem. A*, vol. 1, pp. 3177–3181, 2013.
23. X. Huang, B. Sun, K. Li, S. Chen, and G. Wang, "Mesoporous graphene paper immobilised sulfur as a flexible electrode for lithium–sulfur batteries," *J. Mater. Chem. A*, vol. 1, pp. 13484–13489, 2013.
24. X. Zhao, C. M. Hayner, M. C. Kung, and H. H. Kung, "Flexible holey graphene paper electrodes with enhanced rate capability for energy storage applications," *ACS Nano*, vol. 5, pp. 8739–8749, 2011.
25. P. Lian, X. Zhu, S. Liang, Z. Li, W. Yang, and H. Wang, Large reversible capacity of high quality graphene sheets as an anode material for lithium-ion batteries," *Electrochim. Acta*, vol. 55, pp. 3909–3914, 2010.
26. W. Zhang, J. Zhu, H. Ang, Y. Zeng, N. Xiao, Y. Gao, W. Liu, H. H. Hng, and Q. Yan, "Binder-free graphene foams for O_2 electrodes of Li–O_2 batteries," *Nanoscale*, vol. 5, pp. 9651–9658, 2013.
27. F. Liu, S. Song, D. Xue, and H. Zhang, "Folded structured graphene paper for high performance electrode materials," *Adv. Mater.*, vol. 24, pp. 1089–1094, 2012.
28. P. Guo, H. Song, and X. Chen, "Hollow graphene oxide spheres self-assembled by W/O emulsion," *J. Mater. Chem.*, vol. 20, pp. 4867–4874, 2010.
29. X. Huang, Z. Yin, S. Wu, X. Qi, Q. He, Q. Zhang, Q. Yan, F. Boey, H. Zhang, "Graphene-based materials: synthesis, characterization, properties, and applications," *Small*, vol. 7, pp. 1876–902, 2011.

30. J. P. Meyers, M. Doyle, R. M. Darling, and J. Newman, "The Impedance Response of a Porous Electrode Composed of Intercalation Particles," *J. Electrochem. Soc.*, vol. 147, p. 2930, 2000.
31. J. Euler, and W. Nonnenmacher, "Current distribution in porous electrodes," *Electrochim. Acta*, vol. 2, p. 268, 1960.
32. S. Yun, Y. C. Lee, and H. S. Park, "Phase-controlled iron oxide nanobox deposited on hierarchically structured graphene networks for lithium ion storage and photocatalysis," *Sci. Rep.*, vol. 6, pp. 19959, 2016.
33. G. Zhou, E. Paek, G. S. Hwang, and A. Manthiram, *Adv. Energy Mater.*, vol. 6, p. 1501355, 2016.
34. F.-C. Liu, W.-M. Liu, M.-H. Zhan, Z.-W. Fu, and H. Li, "High-Performance Lithium-Sulfur Batteries with a Self-Supported, 3D Li2S-Doped Graphene Aerogel Cathodes," *Energy Environ. Sci.*, vol. 4, pp. 1261–1264, 2011.
35. X. Wang, Z. Wang, and L. Chen, "Reduced graphene oxide film as a shuttle-inhibiting interlayer in a lithium–sulfur battery," *J. Power Sources*, vol. 242, pp. 65–69, 2013.
36. H. Sun, Z. Xu, and C. Gao, "Multifunctional, ultra-flyweight, synergistically assembled carbon aerogels," *Adv. Mater.*, vol. 25, no. 18, pp. 2554–2560, 2013.
37. Z.-Y. Sui, Q.-H. Meng, J.-T. Li, J.-H. Zhu, Y. Cui, and B.-H. Han, "High surface area porous carbons produced by steam activation of graphene aerogels," *J. Mater. Chem. A*, vol. 2, no. 25, p. 9891e9898, 2014.
38. H. Song, H. X. Wang, Z. Lin, Y. Linwei, X. Jiang, Z. Yu, X. Jun, L. Pan, M. Zheng, Y. Shi, and K. Chen, "Hierarchical nano-branched c-Si/SnO2 nanowires for high areal capacity and stable lithium-ion battery," *Nano Energy*, vol. 19, p. 511, 2016.
39. J. Jin, S. Z. Huang, J. Liu, Y. Li, L. H. Chen, Y. Yu, H. E. Wang, C. P. Grey, and B. L. Su, "Phases Hybriding and Hierarchical Structuring of Mesoporous TiO 2 Nanowire Bundles for High-Rate and High-Capacity Lithium Batteries," *Adv. Sci.*, vol. 2, p. 1500070, 2015.
40. Y. Nishi, "Lithium ion secondary batteries; past 10 years and the future," *J. Power Sources*, vol. 100, pp. 101–106, 2011.
41. N. A. Kumar, R. R. Gaddam, S. R. Varanasi, D. Yang, S. K. Bhatia, and X. S. Zhao, "Sodium ion storage in reduced graphene oxide," *Electrochim. Acta*, vol. 214, pp. 319–325, 2016.
42. X. Xin, X. Zhou, J. Wu, X. Yao, and Z. Liu, "Scalable Synthesis of TiO2/Graphene Nanostructured Composite with High-Rate Performance for Lithium Ion Batteries," *ACS Nano*, vol. 6, pp. 11035–11043, 2012.
43. M. Dubarry, C. Truchot, A. Devie, B. Y. Liaw, K. Gering, S. Sazhin, D. Jamison, and C. Michelbacher, "Evaluation of commercial lithium-ion cells based on composite positive electrode for plug-in hybrid electric vehicle (PHEV) applications," *J. Electrochem. Soc.*, vol. 162, pp. A1787–A1792, 2015.
44. A. J. Smith, S. R. Smith, T. Byrne, J. C. Burns, and J. R. Dahn, "Synergies in Blended LiMn2O4 and Li[Ni1/3Mn1/3Co1/3]O2 Positive Electrodes," *J. Electrochem. Soc.*, vol. 159, pp. A1696–A1701, 2012.
45. B. Qu, C. Ma, G. Ji, C. Xu, J. Xu, Y. S. Meng, T. Wang, and J. Y. Lee, "Layered SnS2-Reduced Graphene Oxide Composite – A High-Capacity, High-Rate, and Long-Cycle Life Sodium-Ion Battery Anode Material," *Adv. Mater.*, vol. 26, pp. 3854–3859, 2014.

6 Current and Future Perspective

Harsh A. Chaliyawala, Kashinath Lellala, Govind Gupta and Indrajit Mukhopadhyay

The development of graphene by using low-cost and naturally available camphor as a carbon source has led to further expanding its reliability into various applications in energy conversion and storage devices. A homogenous island growth mechanism has been identified in the present work to achieve mono/bilayer graphene sheets using natural camphor by a very facile APCVD technique. The existence of the formation of graphene sheet has been determined by an intense 2D band at ~2700 cm^{-1} and G band at 1590 cm^{-1}, respectively, for slow growth process at the rate of ~20 mg/min with 3.5 mg camphor kept at a constant camphor-to-substrate distance.

Since the last decade, there has been an immense interest in combining camphor-based graphene with Si to develop a Schottky junction, which has shown a great potential in solar cell, sensor and photodetector applications. Camphor-based graphene can also be used in various optoelectronic applications where a fast switching action is required at very low power signals in the field of terahertz, optical communication, infrared imaging, and so on. Despite many advantages of the Schottky junction–based devices, their practical implications are still endless and complex in nature. For photovoltaic applications, the photocurrent efficiency (PCE) is very low as compared to commercial Si p–n junction solar cells. The zero band gap of graphene restricts its practical application to some extent. The modification of the graphene surface or structure by various physical and chemical processes can change the graphene Fermi level and thus can regulate graphene photoelectric properties. In addition, the effective active area of graphene/Si-based solar cells is relatively very small, which hinders the electrical performance of the device. Camphor based graphene synthesis can promise large-scale integration of graphene sheets and can overcome the barrier for practical applications. The future work in this direction should be carried out to address two major concerns (i) reduction of sheet resistance and (ii) large-scale production of graphene by an easy and scalable approach, to serve up to the desirable level for practical applications. With all the consequences, the present work demonstrates the highest responsivity of 12.5 A/W for Gr/30 min SiNWAs junction at a very low power signal of 33 μW. Therefore, a facile and low-cost development of photodetectors makes it a favorable candidate for high-performance SiNWAs based NIRPDs.

Moreover, fabrication of homogenous island and defect-free carbon structuring of graphene films from camphor was successfully investigated for LIB's and can be used for other energy-related applications. Further, controlling the growth and island formation in graphene layers such as single- and few-layer sheets can enhance the physicochemical properties that can exhibit excellent LIB's performance. Developing low resistance and high conductivity layers of graphene sheets with high carbon bonding and defect-free sp^2 bonding is highly desirable for energy-related applications. Apart from graphene layers, synthesis and processing of nitrogen- or boron-doped graphene sheets can be further examined for better performance in energy-related applications.

Index

N

O

P

Q

R

S

T

W